Photochemical Processes in Continuous-Flow Reactors

From Engineering Principles to Chemical Applications

Photochemical Processes in Continuous-Flow Reactors

From Engineering Principles to Chemical Applications

Editor

Timothy Noël

Eindhoven University of Technology, The Netherlands

World Scientific

EW JERSEY • LONDON • SINGAPORE • BEIJING • SHANGHAI • HONG KONG • TAIPEI • CHENNAI • TOKYO

Published by

World Scientific Publishing Europe Ltd.

57 Shelton Street, Covent Garden, London WC2H 9HE

Head office: 5 Toh Tuck Link, Singapore 596224

USA office: 27 Warren Street, Suite 401-402, Hackensack, NJ 07601

Library of Congress Cataloging-in-Publication Data

Names: Noël, Timothy, 1982– editor.

Title: Photochemical processes in continuous-flow reactors :
from engineering principles to chemical applications /
by editor Timothy Noël (The Netherlands Eindhoven Univ. of Technology).

Description: New Jersey : World Scientific, 2017. | Includes bibliographical references.

Identifiers: LCCN 2016038444 | ISBN 9781786342188 (hc : alk. paper)

Subjects: LCSH: Photocatalysis. | Chemical reactors.

Classification: LCC QD716.P45 P437 2017 | DDC 660/.2832--dc23

LC record available at https://lccn.loc.gov/2016038444

British Library Cataloguing-in-Publication Data

A catalogue record for this book is available from the British Library.

Desk Editors: V. Vishnu Mohan/Mary Simpson

Typeset by Stallion Press
Email: enquiries@stallionpress.com

Preface

This book had its genesis after my group wrote a review about photochemical transformations in microreactors. The review was picked up by Merlin Fox from *World Scientific Publishing* and he approached me with the idea to write a book about the topic. Given the popularity of photochemistry in flow and the lack of any other book in this field, I was immediately enthusiastic. The rest is history and the final result you have in your hands.

The book aims to give an overview of the important engineering aspects associated with photochemical processes in flow and to provide some relevant examples in the field of organic chemistry and material science. Consequently, it bridges the field of fundamental chemical engineering and organic synthesis. It is my hope that the knowledge gathered in this book will stimulate your imagination for your own research in photochemistry and flow reactors.

I would like to thank all the colleagues and friends who contributed to this volume. Without their help and speedy writing, the book would never have been possible! I also would like to thank the publishing team of *World Scientific Publishing* and, more specifically, Merlin Fox and Mary Simpson who were always prepared to help wherever/whenever necessary.

Enjoy! And may this book pique your imagination for your own research!

Timothy Noël
Eindhoven University of Technology
July 2016

About the Editor

Timothy Noël, born 1982 in Aalst (Belgium), received in 2004 his MSc degree (Industrial Chemical Engineering) from the KaHo Sint-Lieven in Ghent. He then moved to Ghent University to obtain a PhD at the Laboratory for Organic and Bioorganic Synthesis under the supervision of Professor Johan Van der Eycken (2005–2009). The title of his PhD manuscript is "Synthesis and application of chiral dienes and chiral imidates for asymmetric transition metal catalysis". Next, he moved to Massachusetts Institute of Technology (MIT) as a Fulbright Postdoctoral Fellow with Professor Stephen L. Buchwald. At MIT, he worked on the development of new continuous-flow methods for cross-coupling chemistry at the MIT-Novartis Center for Continuous Manufacturing. In 2011, he accepted a position as an assistant professor in the research group of Professor Volker Hessel at Eindhoven University of Technology. His research interest are flow chemistry, homogeneous and heterogeneous catalysis and organic synthesis.

He received in 2011 the Incentive Award for Young Researchers from the Comité de Gestion du Bulletin des Sociétés Chimiques Belges, in 2012 a VENI award from NWO and he was also finalist of the European Young Chemist Award 2012. In 2013, he received a Marie Curie Career Integration Grant from the European Union. Since 2015, he coordinates the Marie Skłodowska-Curie ETN program Photo4Future on the development of photoredox catalysis in photomicroreactors (www.photo4future.com). In 2015, he obtained a prestigious VIDI award from NWO. And in 2016, he received the Thieme Chemistry Journal Award. He serves as an associate editor for *Journal of Flow Chemistry*.

Website: www.NoelResearchGroup.com

Twitter: @NoelGroupTUE

Contents

Chapter 1

Design Consideration
of Continuous-Flow Photoreactors

Stefano Protti, Davide Ravelli and Maurizio Fagnoni*

PhotoGreen Lab, Department of Chemistry
Viale Taramelli 12, Pavia 27100, Italy
*fagnoni@unipv.it

1.1 Introduction

In modern synthesis, a typical multistage procedure involves the isolation and purification of intermediate products after the completion of each step. Obviously, this approach is not only wasteful but also time as well as energy consuming. An ideal model to confront with are the single-cell multistep biosynthetic pathways found in nature, where a *continuous flow* operates from the starting substrates to the desired target. In recent years, a number of innovative strategies have been developed to improve this highly labor- and resource-intensive work by streamlining multistep syntheses, and the most popular approach is represented by the use of continuous-flow techniques. In this case, multiple synthetic steps are combined into a single, continuous and uninterrupted reactor network, thereby avoiding work up and isolation of the intermediate products (for recent reviews on continuous-flow chemistry, see Refs. 1–10). Flow-through processes

1

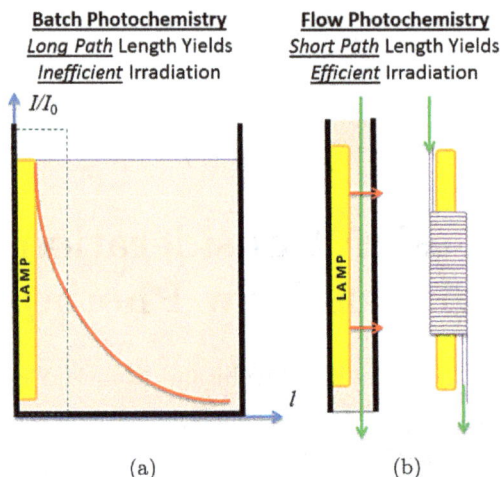

Batch Photochemistry
Long Path Length Yields
Inefficient Irradiation

Flow Photochemistry
Short Path Length Yields
Efficient Irradiation

(a)   (b)

Fig. 1.1. Continuous-flow photochemistry. As shown in the figure (a), the intensity of light (red line) decreases logarithmically with respect to path length, making flow chemical systems more efficient (b). (Reprinted with permission from Ref. 12. Copyright © 2014 The Chemical Society of Japan and Wiley-VCH Verlag GmbH & Co. KGaA, Weinheim.)

are often considered in the scale-up of a synthesis and are generally more satisfactory in terms of yields, reaction times and safety hazards. A notable case, where the advantages of this approach are apparent, is photochemistry, that still has to work hard to be considered as a routine synthetic tool. The reasons for the distrust of mainstream chemists towards photochemical approaches lie in several contributing factors, namely the need of specific know-how and equipment (lamps and unusual glassware, among the others), safety issues when damaging UV radiation and Hg lamps operating at high temperatures are employed and, lastly, the difficulties in scale-up.[11] As for the last point, the key issue is light penetration into the solution, which decreases logarithmically with the path length of the solution, according to the Lambert–Beer law (see Figure 1.1).

For these reasons, large-scale photoreactions in batch reactors are generally inefficient and sluggish, and often result in decomposition of the desired product(s) due to over-irradiation, the Toray process for the production of ε-caprolactam and the Dragoco rose oxide synthesis being the only notable exceptions.[13] These issues have motivated photochemists to develop falling-film and other loops reactors, for which scale-up is more feasible, though in this case a continuous recycle of the reactants solution is required to achieve complete conversion.[14] Notably, loop facilities have been intensively employed when concentrated sunlight was used as the light source.[15,16]

On contrast, when moving to flow photochemical technologies, the reduced size of the reaction channels allows for an efficient light absorption by the substrate, along with a more precise control of reaction time, temperature, pressure and mixing. Furthermore, since the product is continuously removed from the irradiation window, degradation resulting from over-exposure of the desired material is avoided. The advantages commonly consist in improved selectivity, conversion and yield, as well as in the reduction of the reaction time and in a minimization of waste production, thus making flow reactors advantageous for industrial and "green" processes.[17]

Historically, rudimentary flow techniques for photochemical applications have been only sparsely reported in the literature, including the use of a spiral glass reactor in vitamin D synthesis (1959),[18] the use of coiled teflon tubing as a gas-phase reactor for the synthesis of methyl chloride (1971)[19] and a procedure for the removal of photolabile protecting groups in peptide synthesis (1972).[20] However, only in early twenty-first century the use of these apparatuses has become intensive. As highlighted by Noël and coworkers,[21] photons are "reactants" with characteristics different from mainstream chemicals and, accordingly, several factors must be taken into account in the design of a flow photochemical reactor, along with the usual mass and energy balances. The assembling of a photochemical flow setup involves a detailed analysis of all the "actors" involved, namely the reactor (with particular focus on its geometry), the light source, the materials and the chemicals (both solvents and reactants) employed. All of these factors are briefly analyzed in the following to give to practitioners some guidelines to maximize the performance of their photochemical reactions in flow.

1.2 Type and Geometry of the Reactors

The reactors so far available can be divided according to the thickness of the channels where the reaction mixture is circulated, and the corresponding flow rate. Roughly, mesoreactors are characterized by reaction channels with an optical path >0.5 mm, with the flow rate typically exceeding $1\,\mathrm{mL\,min^{-1}}$. On contrast, microflow reactors generally consist of fabricated microchannels usually less than 0.5 mm in thickness that are milled or etched into a planar surface, ranging from bespoke "lab-on-a-chip" designs to highly engineered glass and metal systems (for reviews on microflow photochemistry, see Refs. 11, 17 and 22–28).

In any case, the geometry of the reactor is crucial since the photon transport from the light source to the reaction medium has to be maximized. Irradiation inhomogeneity (e.g. when performing heterogeneous reactions) can cause important local variations in reaction conditions, in turn lowering yield and selectivity. In this case, computational analyses gave a fundamental contribution to address this issue.[29] Chapter 2 will give more insights into this topic, describing radiation and mass transport phenomena in flow photoreactors. Figure 1.2 collects the reactor geometries mainly used in synthetic applications.

Typically, two main approaches have been proposed for assembling a flow photoreactor. In most cases, the reactor is built *around* the lamp (see the case of the annular reactor, Figure 1.2(a)) in order to maximize the capture of photons and hence to increase the productivity (internal irradiation). Several examples made use of the reactor described in Figure 1.3(a),[11] where the solution is pumped in a UV-transparent tubing (see below) wrapped around the light source. These reactors can be easily constructed, even by a novice, using cheap and readily available materials. On the other hand, irradiation can take place outside the reactor apparatus (external irradiation). This is the case of parallel plate reactors (Figure 1.2(b)) and cylindrical reactors (Figures 1.2(c) and 1.2(d)) and these are particularly suited for lab-on-a-chip microreactors, where the solution is irradiated by a light source placed in front of the reactor (Figure 1.3(b)).

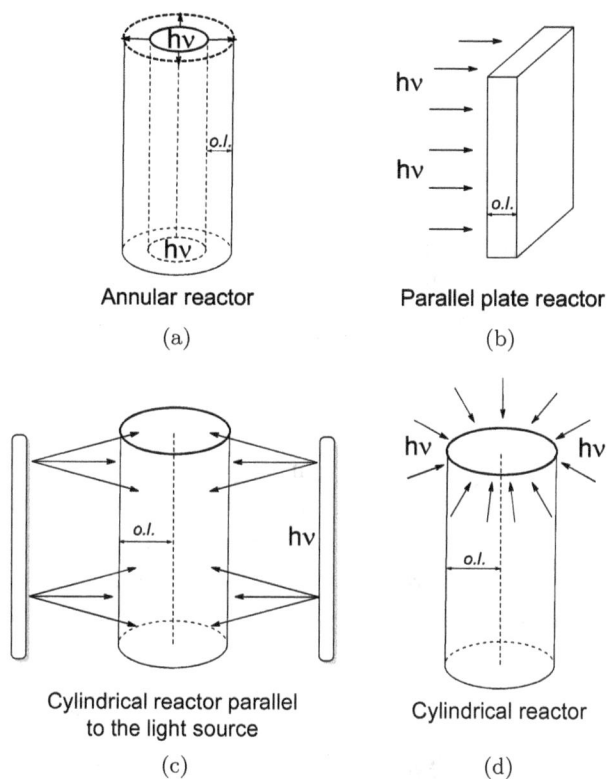

Annular reactor

(a)

Parallel plate reactor

(b)

Cylindrical reactor parallel
to the light source

(c)

Cylindrical reactor

(d)

Fig. 1.2. Reactor geometries with simplified treatments of light sources. *o.l.* = optical length.

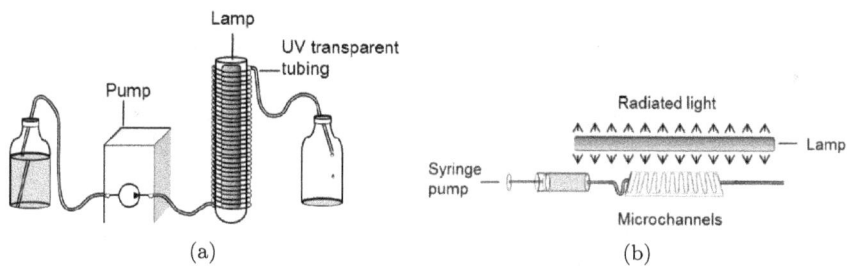

(a)

(b)

Fig. 1.3. Reactors using internal (a) and external (b) irradiation. (Reprinted from Ref. 11 Creative Commons License.)

1.3 Light Sources

In photochemical synthesis, the choice of the light source is subjected to the achievement of two main conditions, viz.:

(a) The wavelength range emitted by the lamp must overlap with the absorption spectrum of the photoactive reactant(s).
(b) Nothing (including the wall of the apparatus) must interfere with the photon before the absorption by the reactant(s).

The most commonly used light sources are summarized in the following (Figure 1.4).

High/medium pressure mercury lamps. These are supplied as small ampoules (from 3 to 15 cm length, depending on the power, that usually ranges from 100 to 1000 W), equipped with a power supply (Figure 1.4(a)). Their emission consists of a range of lines (including 254, 365, 436, 546 and 578 nm) over a continuum background. These light sources develop a considerable amount of heat and running

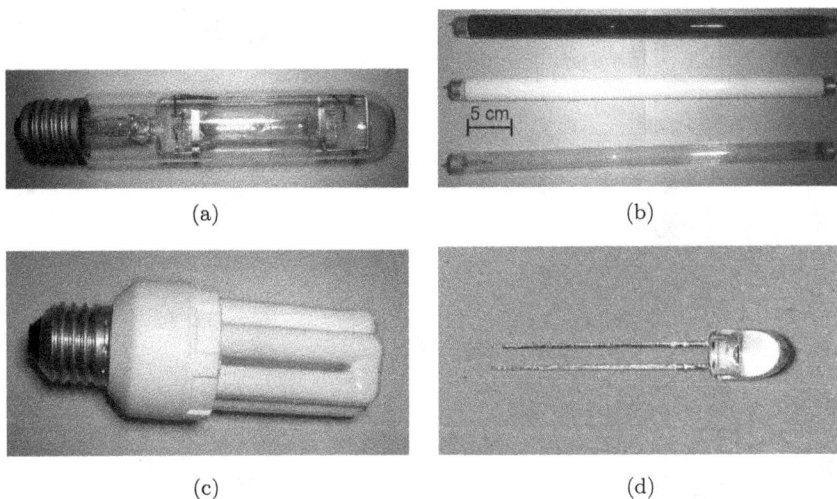

(a)　　　　　　　　　　　(b)

(c)　　　　　　　　　　　(d)

Fig. 1.4. Most commonly used light sources in photochemistry. (a) Bulb of a medium pressure mercury lamp. (b) (Phosphor-coated) low pressure mercury lamps. (c) Compact fluorescent lamp (CFL). (d) Light emitting diode (LED). (Reprinted with permission from Ref. 30. Copyright © 2013 John Wiley & Sons, Inc.)

tap water is sufficient to maintain the temperature at about 20°C. Due to their technical (geometric) characteristics, the elective use of Hg lamps is into immersion-well apparatuses equipped with a water cooling system.

Low pressure Hg arcs. Low pressure (10^{-5} atm under operating conditions) germicidal lamps or mercury resonance lamps are the most widely used light sources in photochemistry. These are supplied as quartz tubes of various lengths (20–60 cm), for which more than 80% of the emission occurs at 254 nm. In any case, the lamp emission can be changed by means of a coating made of a phosphor that absorbs the almost monochromatic Hg radiation and emits a less energetic wavelength (*ca.* 305, 365, 450 nm, Figure 1.4(b)).

These lamps are more long lived (>10000 h) and less energy consuming than the corresponding high/medium pressure mercury lamps. Another important variant is that of *compact fluorescent lamps (CFLs)*. These are energy-saving Hg containing fluorescent lamps, which emit "white" light from a mix of phosphors inside the bulb, with a spectral power distribution that depends on the actual phosphors present (Figure 1.4(c)). The rated service life of these lamps ranges from 6000 to 15000 h, that is almost one order of magnitude higher than that of standard incandescent lamps. A lighting efficiency ranging from 7% to 10% can be achieved in this case.

Wood's light, or simply ultraviolet light, is a variant of such lamps, that emits long-wave ultraviolet light (UV-A) and a minor amount of visible light. These lamps are made of a glass containing *ca.* 9% nickel oxide (giving the typical violet color to the material) absorbing most of the visible light, while allowing UV transmission so that the lamp assumes a dim violet glow when operating.

Light emitting diodes (LEDs) consist of a semiconductor material that emits an incoherent electroluminescence (Figure 1.4(d)). These small devices are fitted with an optics capable to shape and focus the emission and they are available in a large variety of almost monochromatic emitting types (actually over a narrow range, typically 10–20 nm). The conversion of the input electric power into light depends on the output wavelength, being rather poor when emitting UV light. However, LEDs that emit visible light are quite inexpensive and long lived ($>10^4$ h).

Solar light is an environmental friendly alternative, but is characterized by a low density, scarce reproducibility and a limited UV component.[31] However, this light source has been successfully applied to the batch synthesis of several synthetic targets, whereas its use in flow systems is still limited (see below). In alternative, *high-pressure xenon* arc lamps, which emission spectrum mimics solar emission, can be conveniently adopted.

1.4 Materials and Solvents

As hinted above, the materials constituting the reactor must be characterized by a high light transmission and a minimized scattering at the desired wavelength. Moreover, they must be inert to the chemicals used, including reactants and solvents. Accordingly, this greatly limits the range of materials for reactor fabrication. In the following, the most commonly used materials have been reported, along with their most important properties.

Quartz has a wavelength cut-off below 170 nm and allows for the transmission of both UV and visible light. Other silica-based materials are: Vycor (>220 nm transmission), Corex (>260 nm) and Pyrex (>275 nm). All of them have been used in assembling both meso and microphotoreactors. However, whereas glass is widely used for batch photochemistry, the poor workability of glass-based materials limits their application in the assembling of flow reactors. Furthermore, they are prone to undergo erosion under strong alkaline conditions and at high temperature.

Polymeric materials. Several tubings made of polymeric materials have been investigated for the realization of flow photochemical setups. Polymethylmethacrylate (PMMA) and polydimethylsiloxane (PDMS) are characterized by a low cost and an efficient light transmission (above 300 nm in the case of PMMA), but the tendency to swell in the presence of organic solvents limits their practical use

Wavelength

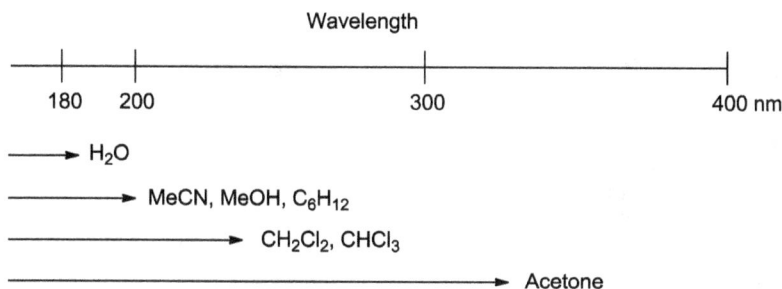

Fig. 1.5. Absorption properties of commonly employed solvents in photochemistry. The arrows indicate the region where the solvent absorbs.

in synthesis. By contrast, perfluoroalkoxyalkanes (PFA, which transmission is up to 96% and 91% for visible and UV light, respectively) and fluorinated ethylene propylene (FEP) are preferred, since they exhibit high flexibility and resistance to relatively high pressures, strong acids and bases, as well as to swell phenomena. These properties make them the election polymers in the preparation of flow photochemical reactors, independently from the chosen scale. In the case of PFA, a high purity material is required, in order to avoid contamination of the desired product (e.g. by plasticizers present in the polymer).[32]

As for the choice of the solvent, it must be obviously transparent to the wavelength used, as in all photochemical processes, to avoid quenching of the desired reaction. A resume of the most important absorption properties of commonly adopted solvents in photochemistry is shown in Figure 1.5.[33]

In this chapter, selected examples of flow processes will be shown according to the irradiation mode (internal or external), highlighting in each case, the geometry employed, the light source used and the material used for assembling the photoreactor.

1.5 Its All Around You: Flow Photoreactors with Internal Irradiation

Several examples belonging to this class took inspiration from the photoreactor assembled by the group of Prof. Booker-Milburn in

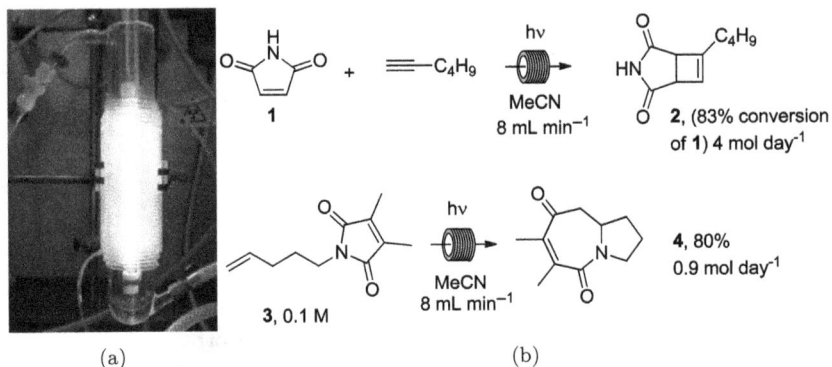

Fig. 1.6. The Booker-Milburn type flow photoreactor (a). Photochemical synthesis of cyclobutene **2** and azepine **4** under flow conditions (b). (Reprinted from Ref. 11. Creative Commons License.)

2005 (see Figure 1.6(a)).[34] This continuous-flow photochemical system consisted of a medium pressure Hg lamp with a cooling well, around which an FEP tubing was wrapped (up to five layers depending on the diameter of the tubing). In the optimized system, an FEP tubing with an optical path length of 2.7 mm was chosen due to its resistance to damage and its capability to hold a considerable volume of solution without generating significant back-pressures, even at high flow rates (up to 10 mL min^{-1} by means of an HPLC pump). The advantages of such an approach are apparent, since the large surface-to-volume ratio ensures an optimized irradiation of the solution. Both intramolecular and intermolecular cyclizations have been investigated, namely the synthesis of cyclobutene **2** (via the [2 + 2] photocycloaddition of maleimide **1** and 1-hexyne)[34] and the [5 + 2] cycloaddition of 3,4-dimethyl-1-pent-4-enylpyrrole-2,5-dione **3** to the bicyclic azepine **4**.[34] Both products have been obtained in high yields, with the productivity exceeding 500 g (for **2**) and 150 g (for **4**) of product per day (Figure 1.6(b)). A similar [5 + 2] intramolecular cycloaddition was carried out in the multistep, protective group-free synthesis of alkaloid (±)-neostenine that took place in 9.5% yield over 14 steps.[35]

The seminal work described above served as the archetype for the development of a plethora of related reactors employing medium

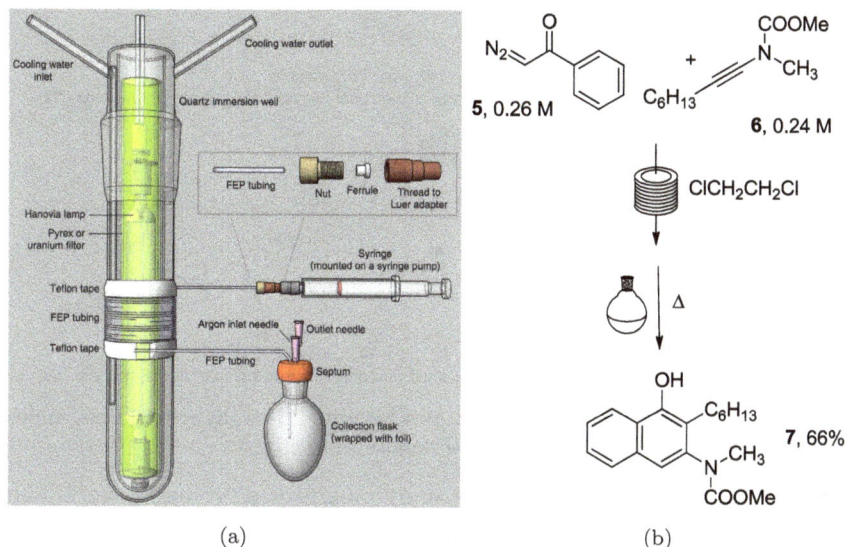

Fig. 1.7. Photochemical benzannulation for the synthesis of substituted aromatic alcohols (e.g. **7**) under continuous-flow conditions. (Reprinted with permission from Ref. 36. Copyright © 2013 American Chemical Society.)

pressure Hg vapor lamps. Since these light sources are known to generate much heat during process illumination, the temperature of the reactor at tubing surface was usually controlled by a thermometer affixed to the coils. Furthermore, materials belonging to different classes have been employed for the immersion well apparatus. In most cases, Vycor was used for allowing to use light sources with an emission above 220 nm. In a recent example, Willumstad *et al.* described the synthesis of differently substituted phenols and naphthols (e.g. **7**) phototriggered by the Wolff rearrangement in diazoketone **5** followed by [2 + 2] cycloaddition with ynamide **6** (Figure 1.7(b)).[36]

In this case, the FEP tubing (1.59 mm o.d.; 0.76 mm i.d.; up to 1340 cm length for an overall volume of 6.1 mL) was coiled around a quartz immersion well. The bottom end of the tubing was connected through a rubber septum to a pear-shaped flask equipped with an argon inlet and an outlet needle. Irradiations were carried out in most cases through a Pyrex filter (Figure 1.7(a)). In a typical process, a deaerated 1,2-dichloroethane solution containing **5** and **6** was

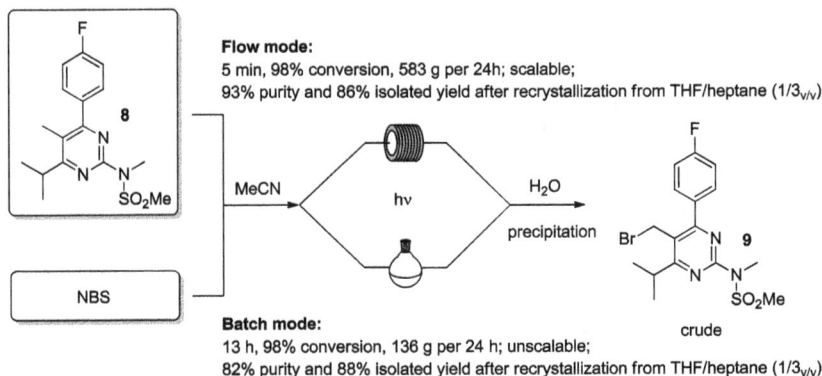

Scheme 1.1. Benzylic bromination of substituted 5-methylpyrimidines under batch and flow conditions. NBS = *N*-bromosuccinimide.

circulated in the flow photoreactor by means of a syringe pump. The resulting mixture was then collected in the pear shaped flask and refluxed for 1.5 h to complete the process (Figure 1.7). The reaction took place in a yield comparable to that obtained in the batch reactor, but the residence time was markedly shortened (from 2.5 h down to *ca.* 20 min), with a typical throughput of 22 mmol day^{-1}.[36]

An 18-mL flow reactor consisting of a 150-W medium-pressure Hg lamp was used to improve both the reaction productivity (from 130 up to 580 g day^{-1} when moving from batch to flow conditions) and the selectivity of the bromination of a 5-methyl substituted pyrimidine (**8**) to **9**, precursor of the cardiovascular therapeutic Rosuvastatin (Scheme 1.1).[37]

Hurevich *et al.*[38] used a Booker-Milburn type reactor sealed in an aluminum box for performing continuous-flow solid phase synthesis (Figure 1.8). In a typical experiment, a polystyrene-based support with photolabile linkers was pre-swollen in dichloromethane and pushed through the FEP tubing using a syringe pump and then irradiated for 30 min. The bottom end of the tubing was connected with a flask fitted with a polyethylene filter to separate the resin from the filtrate. A simple wash with dichloromethane between resin deliveries was sufficient to enable multiple runs in the same reactor. Contrary to the batch reactor, no turn off of the light source was required to remove beads and reagents from the reactor.

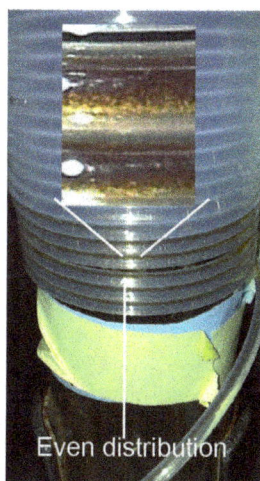

Fig. 1.8. Resin distribution in the flow photoreactor: the constant flow helps to distribute uniformly the resin in the reactor. (Reprinted with permission from Ref. 38. Copyright © 2014 American Chemical Society.)

When irradiation at 254 nm is required, the use of Teflon AF-2400 tubing is mandatory since FEP in part decomposes at this wavelength. As an example, separate solutions of thiol **10** and olefin **11** were delivered by two syringe pumps, mixed with a T-union, and irradiated to perform a thiol-ene reaction. This allowed the preparation of a key building-block (**12**) on the route to the synthesis of carbohydrate-functionalized poly/oligo(amidoamine)s (PAAs; Figure 1.9).[39]

The photochemical flow reactor can be combined with a thermal flow reactor as recently demonstrated in the multistep synthesis of substituted γ-lactones **13** by two consecutive photocatalytic/ reductive reactions.[40] The flow reactor made use of a 125-W Hg lamp and of an Algoflon tubing (poly-tetrafluoroethylene (P-TFE), Figure 1.10). A flow-through back-pressure unit was mounted at the outlet of the reactor to avoid the formation of bubbles during the reaction and to maintain a constant flow rate.[40]

Related tetrabutylammonium decatungstate (TBADT) photo-catalyzed acylations and alkylations of electron-poor olefins by using

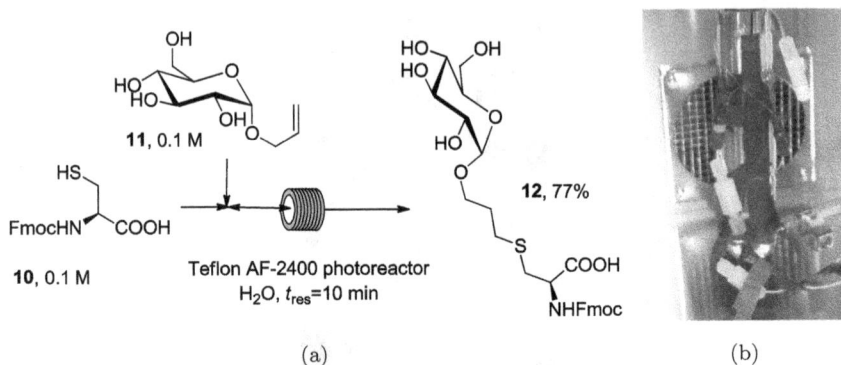

(a)   (b)

Fig. 1.9. Thiol-ene reaction (a) and picture of the continuous-flow reactor (b) used for the synthesis of **12**. (Reprinted with permission from Ref. 39. Copyright © 2013 John Wiley & Sons, Inc.)

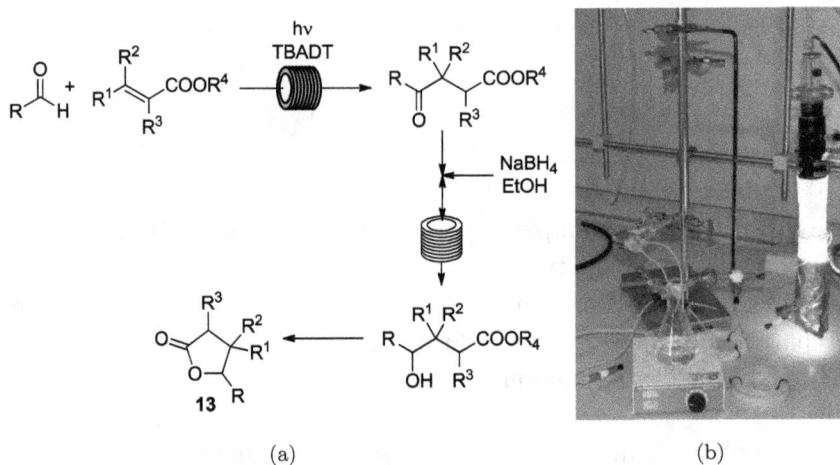

(a)   (b)

Fig. 1.10. Multistep synthesis of γ-lactones **13** (a). Picture of the flow reactor used in the preparation of **13** (b). TBADT = Tetrabutylammonium decatungstate. (Courtesy from PhotoGreen Lab (University of Pavia, Italy).)

a 500-W medium pressure mercury vapor lamp have been likewise devised.[41] In addition, the same 500-W lamp was found convenient for metal-free arylations via photogenerated phenyl cations.[42]

Materials different from fluorinated polymers have also been considered in designing Booker-Milburn-type reactors. A photoreactor

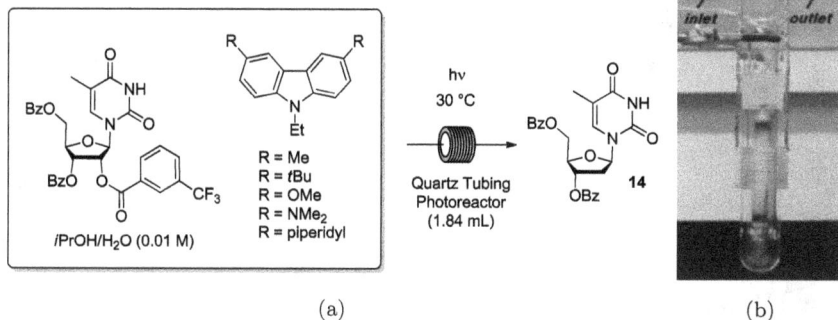

(a)   (b)

Fig. 1.11. Synthesis of 2'-deoxynucleoside **14** (a). Photochemical flow reactor featuring quartz tubing (b). (Reprinted with permission from Ref. 43. Copyright © 2012 The Royal Society of Chemistry.)

having coils of quartz tubing (1 mm i.d., 1.84 mL total volume) was placed approximately at 2 cm from the surface of a 450-W medium pressure Hg lamp and used by Jamison *et al.* for the photosensitized synthesis of 2'-deoxy (**14**) and 2',3'-dideoxynucleosides.[43,44] The apparatus was placed in a Pyrex cylinder coated with aluminum for reflecting the UV radiation, with an estimated efficiency >80%. Water was circulated in the vessel with the aid of a water pump immersed in a temperature-controlled bath, thus allowing the reaction temperature to be carefully set (in 0–50°C range; Figure 1.11).

A different flow photoreactor, consisting in a single layer of FEP tubing coiled around a quartz tube, has been likewise described (Figure 1.12). The system made use of a single germicidal compact fluorescent 36-W PL-L lamp inserted into the quartz tube. The device allowed to perform the synthesis of tricyclic aziridines **15,16** without degradation of the starting pyrroles, that was observed when using high-power medium pressure mercury lamps.[45] Thus, the reactants solution was flowed through the reactor by means of a valveless piston pump. The same reactor has been also used for the synthesis of dipeptides starting from nitrones.[46]

Inexpensive, interchangeable, low-energy 9 W lamps were proposed by Harrowven *et al.*[47] in place of conventional 400–600 W Hg

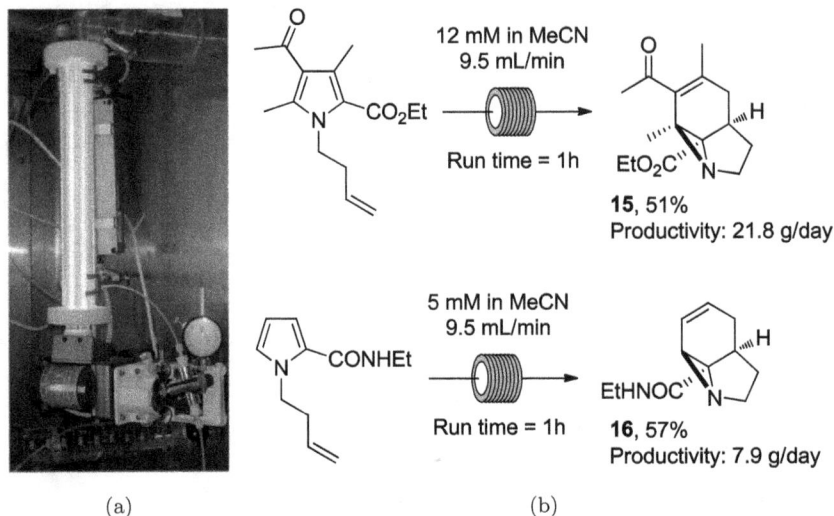

(a)   (b)

Fig. 1.12.   Photocycloaddition reactions of pyrroles. (Reprinted with permission from Ref. 45. Copyright © 2013 John Wiley & Sons, Inc.)

lamps with the aim of making the choice of the wavelength to be used *"as simple as changing a light bulb"*, since commercial low-pressure lamps currently cover a wide range of wavelengths. A thermocouple was incorporated within the coils of the photoreactor to monitor the reaction temperature. The resulting reactor was then used for the preparation of *5H*-furanones from 4-hydroxycyclobutenones in near to quantitative yields, except the case of compound **17** where a 1:1 mixture of compounds **18** and **18′** resulted. By contrast, the corresponding thermal process led to isomeric derivative **18′** (Figure 1.13).

A Wood's light (16 W) could be likewise adopted as the light source in the photoinduced three-component reaction between diazoketones, carboxylic acids and isocyanides for the preparation of 2-acyloxyacrylamides.[48] In alternative, the photochemical reactor could be assembled by surrounding a commercial germicidal compact fluorescent lamp by a PFA tubing (3 mL overall volume; Figure 1.14).[49] This system was then employed for the preparation of β-amino acids **19** from the corresponding α-amino acids via a multistep

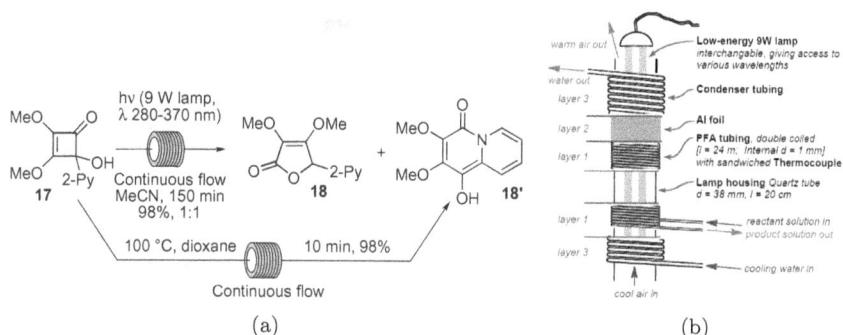

Fig. 1.13. Photochemical rearrangement of (2-pyridyl)-cyclobutenone **17** (a). Scheme of the flow photochemical reactor employed (b). (Reprinted with permission from Ref. 47. Copyright © 2012 John Wiley & Sons, Inc.)

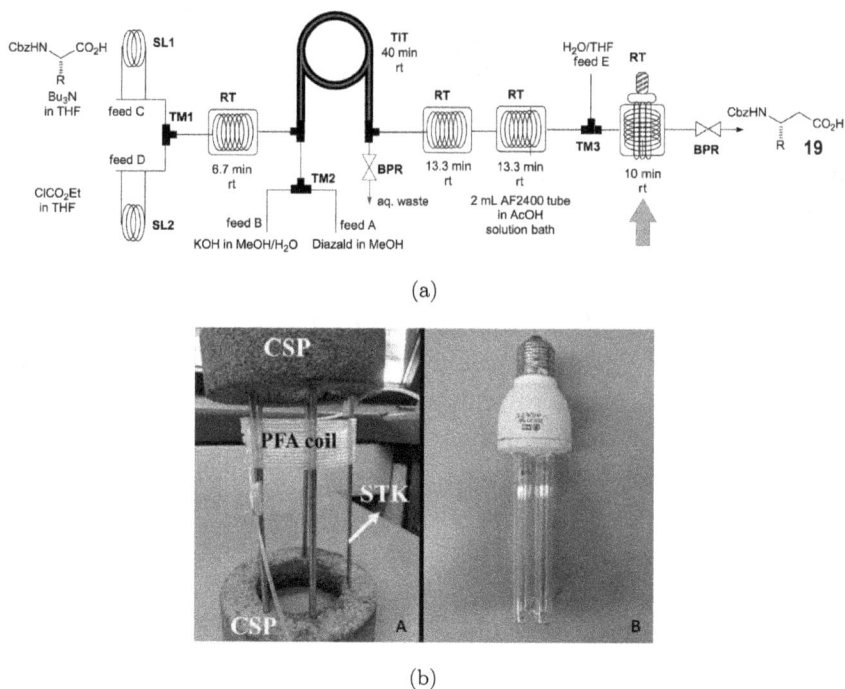

Fig. 1.14. (a) Flow setup for the continuous four-step Arndt–Eistert homologation of α-amino acids (RT = reactor tube). (b) Picture of the photochemical flow reactor indicated by the arrow (A) and the lamp used (B). CSP = flask cork support; STK = metal sticks. (Adapted from Ref. 49. Copyright © 2014 with permission of The Royal Society of Chemistry.)

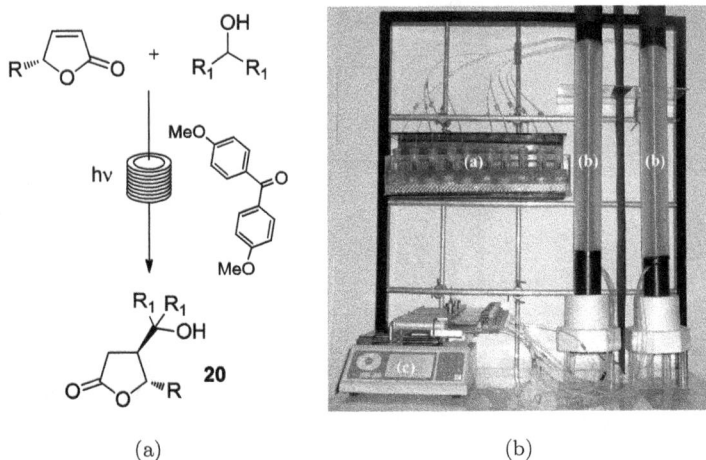

(a)   (b)

Fig. 1.15. Photocatalyzed synthesis of substituted 2(5H)-furanones **20** (a). MμCFR: (a) collection flasks; (b) FEP microcapillaries; (c) 10-syringe pump (b). (Reprinted with permission from Ref. 50. Copyright © 2012 American Chemical Society.)

Arndt-Eistert homologation, the key step being the photoinduced Wolff rearrangement of the *in situ* generated diazoketone.

The internal light approach can also be applied to multimicrocapillary flow reactors (MμCFR), as demonstrated by Yavorsky *et al.*[50] Two bundles of five FEP capillaries (1.6 mm o.d.; 0.8 mm i.d.; 5 mL each) were tightly wrapped around two Pyrex glass columns equipped with 2×8 W UV-A fluorescent tubes ($\lambda_{max} = 365$ nm), in order to achieve a large surface-to-volume ratio (about 2500 m^2/m^3). This rare case of (multi) microreactor with internal irradiation has been applied to the parallel preparation of a library of 2(5H)-furanones **20** (Figure 1.15).[50]

A photoreactor (Figure 1.16) consisting of an FEP tubing coiled around a glass cylinder (28 mL reactor volume) with a black light household CFL (105 W) placed inside the assembled reactor was used by Cantillo *et al.* for the xanthone photocatalyzed fluorination of alkylaromatics. The photoreactor was located inside a GC oven to enable an accurate control of the temperature.

Fluorinated celestolide **21** was thus obtained on a gram scale (88% yield, Figure 1.16)[51] and this approach was later exploited in

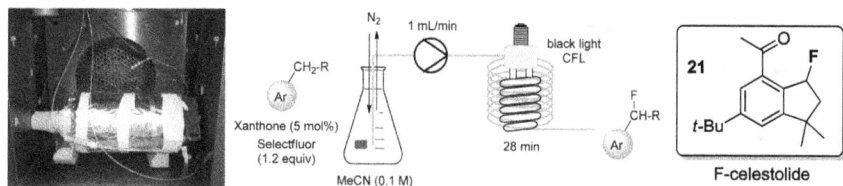

Fig. 1.16. Continuous-flow photoreactor employed for the fluorination of alkylaromatics at the benzylic position. (Reprinted with permission from Ref. 51. Copyright © 2014 American Chemical Society.)

the Eosin photocatalyzed trifluoromethylation[52] and bromination of the same substrate at the benzylic position.[53]

The photocatalyzed (by TBADT) benzylic fluorination of ibuprofen methyl ester (N-fluorobenzenesulfonimide as the fluorine source) was later described by flowing the solution in an FEP tubing wrapped around a black light blue lamp ($\lambda = 365$ nm). Under these conditions, the reaction occurred in only 5 h (24 h in batch) without affecting the overall yield (70%).[54]

Recently, the impressive development of visible light photoredox catalysis[55−60] allowed for the use of low energy demanding and easily available light sources, such as LEDs and CFLs. The latter (26 W) were adopted for the multistep flow synthesis of polycyclic quinoxalines (e.g. **22**), by having recourse to an integrated flow setup consisting of three PFA tubing reactors (R_1, R_2 and R_3 in Figure 1.17). In this case, the photochemical reactor could be refrigerated by immersion in a water-cooling bath. Thus, compound **22** was isolated in a 47% overall yield (*ca.* 78% yield per step; throughput: 6.3 μmol min^{-1}) from the starting aniline (overall residence time of 14.6 min).[61]

A 23-W household fluorescent lamp was used to irradiate the solution of an aldehyde, a bromomalonate, a chiral organocatalyst and a dye (Eosin Y as the photocatalyst) flowing through an FEP tubing wrapped around a glass beaker to perform efficient enantioselective photoredox α-alkylation of aldehydes.[62]

In a related case, a simple apparatus was assembled by placing three strips of 1 W blue LEDs inside a Liebig condenser, while the

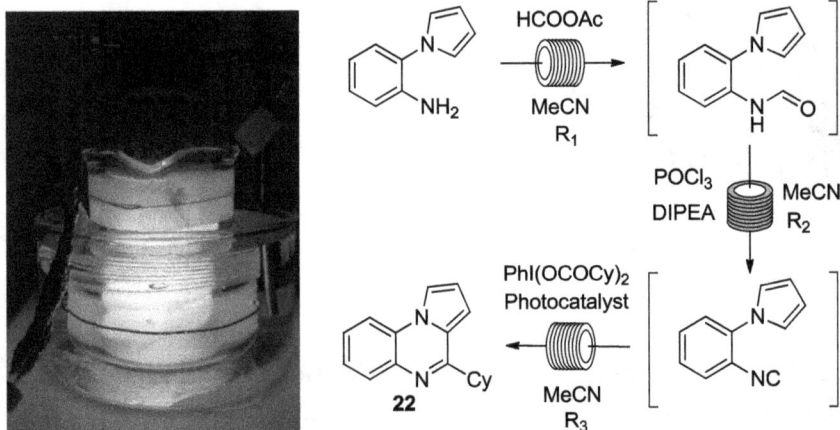

Fig. 1.17. Continuous flow synthesis of pyrrolo[1,2-a]quinoxaline **22**. (Reprinted with permission from Ref. 61. Copyright © 2014 John Wiley & Sons, Inc.)

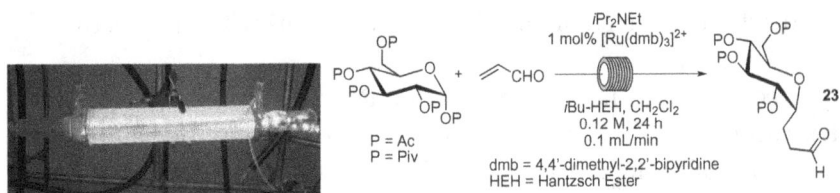

Fig. 1.18. Continuous-flow synthesis of acetyl- (Ac) and pivaloyl- (Piv) protected sugars. (Reprinted with permission from Ref. 63. Copyright © 2012 John Wiley & Sons, Inc.)

reacting solution was pumped into an FEP tubing wrapped around the same condenser. The assembled reactor was then employed in the visible light photoredox catalyzed conjugate addition of O-protected sugars onto acrolein to form **23**, as the first step in the synthesis of different C-glycoaminoacids and C-glycolipids (Figure 1.18).[63]

Flow systems could be easily employed for the continuous lab-scale synthesis of drugs. A photo-Favorskii rearrangement converted an α-chloropropiophenone into ibuprofen in 76% isolated yield in a green medium such as acetone/water (9:1).[64]

The hydrochloric acid liberated in the reaction was scavenged by propylene oxide (2 vol%). The apparatus used is illustrated in

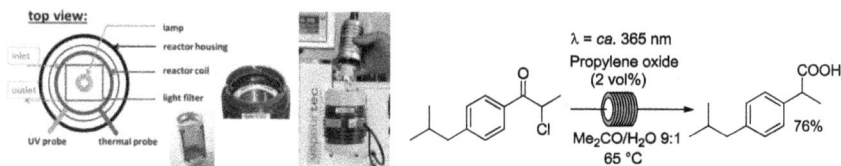

Fig. 1.19.   Continuous-flow synthesis of ibuprofen via photo-Favorskii rearrangement of a chloropropiophenone. (Reprinted from Ref. 64. Copyright © 2016 The Royal Society of Chemistry.)

Figure 1.19 and consists in a photochemical reactor combined with an existing and commercially available flow reactor system. Notably, this setup allowed to perform a real time analysis of the process.[64]

Flow photoreactors were likewise adopted in the photocatalytic degradation of volatile organic compounds (VOCs). As a representative example, a gaseous effluent containing trimethylamine or isovaleraldehyde was introduced in an annular reactor made of two concentric Pyrex tubes, where the inner side of the external tube was covered with the photocatalyst (TiO_2) and irradiated with an 80-W UV lamp placed in the center of the inner tube.[65]

1.6   Flow Photoreactor with External Irradiation

In a recent example, Oelgemoeller and coworkers placed a microcapillary tower in a Rayonet chamber reactor equipped with 8×4 W UV-C lamps. The $[2 + 2]$ cycloaddition described in Figure 1.20 was used as a model reaction.

Compared to batch conditions, the microcapillary unit had a 9-time larger surface-to-volume ratio (more than $3000\,m^2/m^3$), that provided a more efficient light absorption (Figure 1.20). The cycloadduct **24** was obtained in comparable yields by means of both batch and flow reactors, but shorter irradiation times were required in the latter case.[66]

In a further photoredox catalyzed process, a 4.7 mL-photoreactor, consisting of an FEP tubing wrapped around two vertical metal rods, was placed between two 17-W cold white LED lamps. The high photon flux at 400–500 nm matched with the typical metal to ligand charge transfer (MLCT) absorption band of Ruthenium(II)

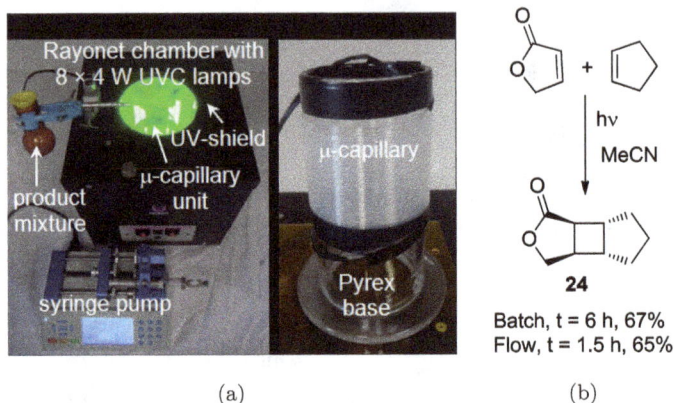

(a)　　　　　　　　　　　　　　　　(b)

Fig. 1.20. Microcapillary reactor. Photochemical setup containing an inserted μ-capillary unit with a close-up of the μ-capillary unit (a). Comparison between batch and flow conditions for the synthesis of cyclobutane rings (b). (Reprinted from Ref. 66. Creative Commons License.)

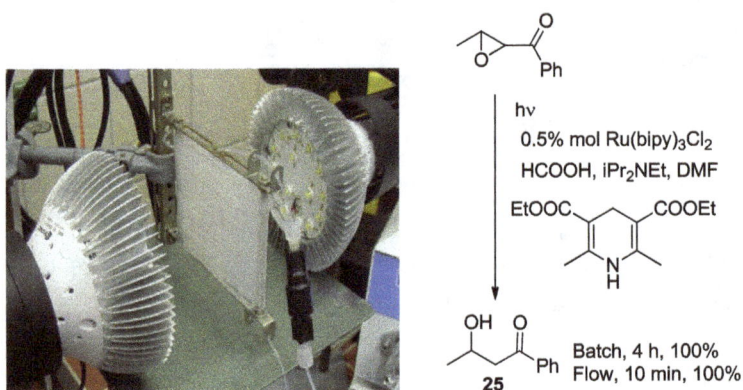

Fig. 1.21. Reactor setup employed for the reductive opening of chalcone-α, β-epoxides. (Reprinted with permission from Ref. 67. Copyright © 2012 The Royal Society of Chemistry.)

complexes (maximum centered around 450 nm) and was exploited by Seeberger *et al.* to perform Ru-photoredox catalyzed transformations (an example in Figure 1.21 for the synthesis of **25**) with a 10- to 50-fold rate enhancement over the corresponding batch conditions and a minimal waste generation.[67]

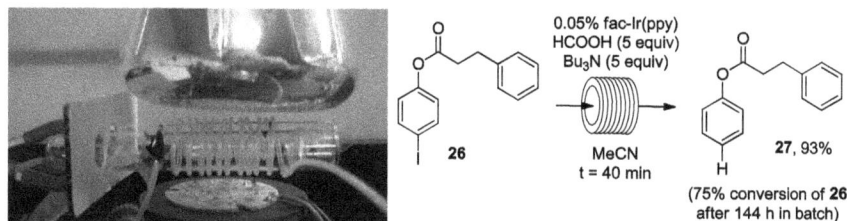

Fig. 1.22. Picture of the photoreactor used in the photocatalyzed reduction of aryl iodides. (Reprinted with permission from Ref. 68. Copyright © 2012 John Wiley & Sons, Inc.)

Similarly, Stephenson and coworkers implemented a wide range of photoredox catalyzed reactions by assembling a flow reactor constituted of a PFA tubing wrapped around two or three glass tubes. The solution was externally irradiated by using seven blue LEDs positioned 2 cm below the coiled tubing. A mirrored Erlenmeyer flask was positioned above the reactor in order to reflect the incident light back onto the tubing.[68] Several processes, including C–C bond formation, one-pot deoxygenation of primary and secondary alcohols[69] and dehalogenation of alkyl, alkenyl and aryl iodides,[70] have been investigated by having recourse to this device, with satisfactory yields and short residence times (down to 1 min). As an example, aryl iodide **26** was reduced to **27** in 88% yield with a 40-min residence time in the presence of 0.05 mol% of an Ir-based catalyst, with a 120-fold improvement compared to the conversion rate obtained under batch conditions (Figure 1.22).

An analogous device was obtained by simply wrapping a PFA capillary tubing around a large transparent disposable syringe, and placing the reactor inside a beaker with an array of blue LEDs coiled in a spiral fashion.[71] A gram-scale trifluoromethylation of 3-methyl indole **28** with gaseous CF$_3$I to give **29** (1.98 g, 9.5 mmol) took place quantitatively in less than 9 h (Figure 1.23). Again, the use of flow conditions allowed to decrease the catalyst loading down to 0.05 mol% without loss of efficiency (75% yield).

Natural antimalarian artemisinin **34**, which is industrially obtained via extraction from the sweet wormwood plant, was

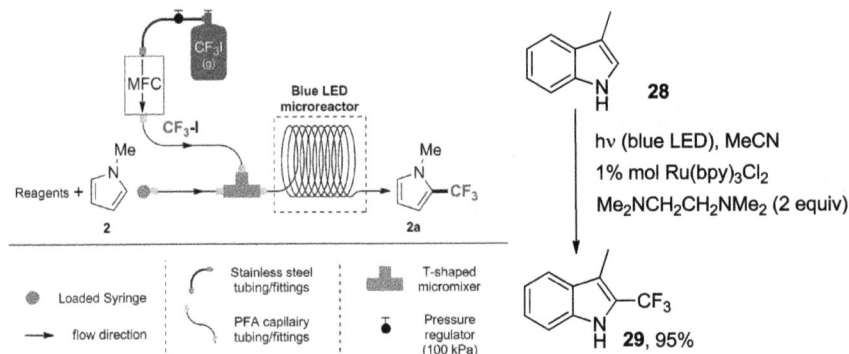

Fig. 1.23. Schematic representation of the continuous-flow setup for the trifluoromethylation of substituted indoles. (Reprinted with permission from Ref. 71. Copyright © 2014 John Wiley & Sons, Inc.)

synthesized on a multigram-scale by Lévesque and coworkers exploiting a flow approach. The key step of the process was the tetraphenyl porphyrin (TPP) photosensitized oxidation of dihydroartemisinic acid **30**. Initially, this step was carried out in a Booker-Milburn photoreactor which provided the allylic hydroperoxide **31** in 75% yield (91% conversion). Subsequently, an acid-induced Hock cleavage/oxygen-induced condensation cascade promoted by one equivalent of trifluoroacetic acid (TFA) furnished the desired product **34** in 39% yield.[72]

Later, the Booker-Milburn reactor was replaced by an FEP tubing wrapped around a polycarbonate plate in four layers. (Figure 1.24(a)) The resulting reactor was immersed in a glycol/water $(3/2_{v/v})$ mixture to control the temperature and irradiated with an LED module (420 nm wavelength) at a distance of 3 cm from the tubing. This device allowed the authors to operate at −20°C (in order to maximize the oxidation yield) and resulted in the production of artemisinin in 65% overall yield (Figure 1.24). In the latter case, 9,10-dicyanoanthracene (DCA) was used as the photosensitizer in the place of TPP.[73] Interestingly, large amounts of **30** were found in the mother liquors deriving from extraction of artemisinin from *Artemisia annua*, thus making this solution usable as the starting mixture.[73]

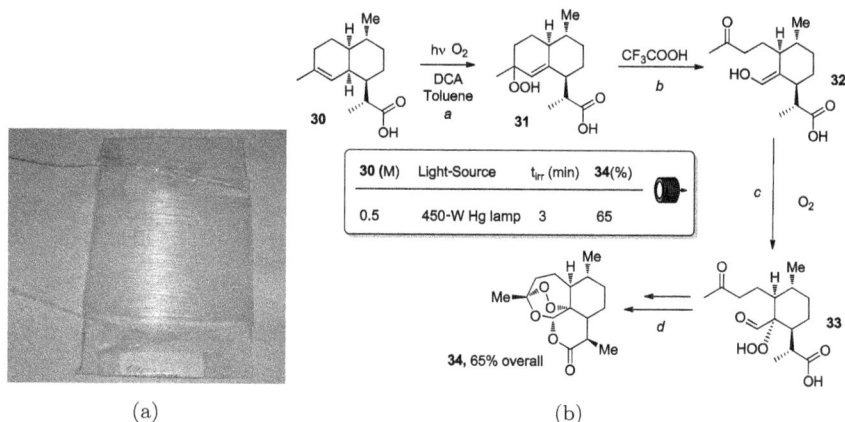

Fig. 1.24. Photoreactor consisting of FEP tubing wrapped around a transparent polycarbonate plate used for the synthesis of Artemisinin **34** (a). The steps involved in the synthesis of **34** (b). (Reprinted with permission from Ref. 73. Copyright © 2013 John Wiley & Sons, Inc.)

Very recently, a one-pot flow route to **34** was developed by using a photoreactor equipped with three banks of LEDs. The reaction was performed on a solution of **30** in an EtOH/H$_2$O mixture in the presence of Ru(bpy)$_3$Cl$_2$ (as the photosensitizer) and H$_2$SO$_4$. The resulting artemisinin was obtained in high purity by simply evaporating EtOH from the reaction mixture. This innovative strategy allowed a decrease of the E-factor value from 4.8 (calculated for the industrial process) down to 1.3 kg kg^{-1}.[74]

The same setup has been employed for the oxidative cyanation of primary and secondary amines to aminonitriles, that occurred at −50°C in the presence of trimethylsilyl cyanide and TPP as the sensitizer.[75] As an example, neopentylamine was converted into the corresponding nitrile in high yield and further elaborated to give *d,l*-*tert*-leucine hydrochloride upon hydrolysis in concentrated HCl (87% yield over three steps).

FEP tubing filled with a mixture of silica gel, glass beads and the heterogeneous photocatalyst graphitic carbon nitride (mpg-C$_3$N$_4$) constituted an ideal continuous-flow photoreactor in which Woźnica *et al.* optimized the metal-free radical synthesis of functionalized

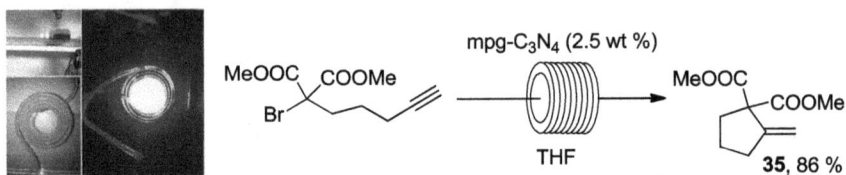

Fig. 1.25. Continuous-flow system for the synthesis of cyclopentanes (e.g. **35**). (Reprinted with permission from Ref. 76. Copyright © 2014 John Wiley & Sons, Inc.)

cyclopentanes (e.g. **35**) from 2-bromo-1,3-dicarbonyl compounds under irradiation with LEDs ($\lambda = 420$–425 nm, Figure 1.25).[76]

The photooxygenation of monoterpenes was carried out in a reactor formed using two T-junctions and a transparent tube-in-tube (TIT) system. The outer layer was a transparent and gas non-permeable polyether ether ketone (PEEK) tube, while the inner layer was a chemically resistant and gas-permeable Teflon tube. The photosensitizer (methylene blue (MB)) and the terpene were mixed at the first T-junction and the resulting solution met oxygen gas at the second one, allowing a high oxygen–liquid contact area. This technological approach showed industrial applications with a daily output 270-fold higher than that in typical batch systems.

Oxidation of ($-$)-citronellol **36** to a mixture of allylic peroxides **37**, **38**, the key step in the industrial production of the fragrance rose oxide, took place almost quantitatively and in a 68% yield by using white LEDs and upon exposure to natural sunlight, respectively (Figure 1.26).[77]

The commercially available Ehrfeld Photoreactor XL with tunable optical path was recently combined with a custom-built bank of four 8 W low-pressure lamps and used in the optimization of the intramolecular cycloaddition of cyclopentenone **39** to tricyclic ketone **40**. Under optimized conditions, 100 g of starting **39** ($0.5\%_{w/v}$ in acetone) were processed in 111 h and afforded **40** with a 48% yield (Scheme 1.2).[78]

The photocatalytic oxidation of arylboronic acids to phenols has been carried out in a high pressure tubular sapphire photoreactor with an upflow arrangement, fitted with three banks of high-power

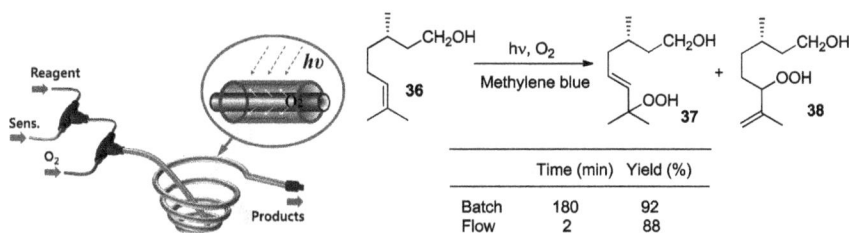

Fig. 1.26. Tube-in-tube reactor (TIT-R) used for the photo-oxygenation of cytronellol **36**. Oxygen gas was injected into a gas-permeable inner tube (AF-2400; 0.8 mm o.d.; 0.6 mm i.d.) and diffused into the liquid phase containing the reagent and the photosensitizer in the transparent outer tube. The liquid phase was held against an LED light. (Reprinted with permission from Ref. 77. Copyright © 2015 The Royal Society of Chemistry.)

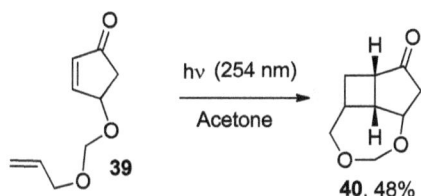

Scheme 1.2. Synthesis of annulated cyclopentanone **40**.

white LEDs (Figure 1.27). The organic feed contained the dissolved substrate in an EtOH/H_2O mixture and the organic photocatalyst (rose bengal), while the reductive quencher was in a flask open to air. The starting solution was pumped using an HPLC piston pump and air was dosed via a Rheodyne dosing valve, while the different parts of the reactor were connected with 1/16″ stainless steel tubing. The photoreactor was filled with 6 mm diameter glass beads to promote mixing and to reduce the optical path length, thus preventing inner filter effects. During the irradiation, the sapphire tube was surrounded by a concentric transparent cooling jacket for temperature control.[79]

Recently, several microflow reactor systems have been designed and some of them can also be found nowadays on the market[a]

---

[a]Main distributors of photochemical microreactors are Future Chemistry, Mikroglas Chemtech and Sigma Aldrich.

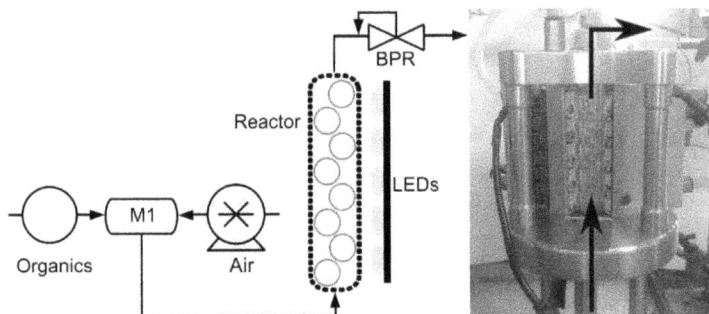

Fig. 1.27. The sapphire tubular photoreactor in upflow configuration. (Reprinted with permission from Ref. 79. Copyright © 2015 The Royal Society of Chemistry.)

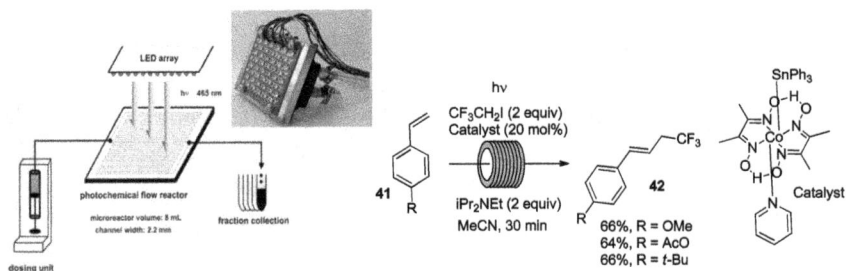

Fig. 1.28. Photochemical apparatus employed for trifluoroethylations in flow. (Reprinted with permission from Ref. 80. Copyright © 2013 American Chemical Society.)

(Figure 1.28).[24] An example is the self-made flow reactor photobox consisting of a single-channel syringe pump, a water-cooled array of 48 high-power LEDs (total radiant flux of *ca.* 37 W) and an 8-mL glass microreactor. The box was lined with silver glass mirrors for optimal reflectivity. Trifluoroalkyl derivatives **42** have been synthesized in a continuous mode via a Co-photocatalyzed functionalization of styrenes **41** with a residence time of 30 min. Notably, the same process in batch required 24 h to reach completion (Figure 1.28).[80]

The continuous-flow synthesis of vitamin D$_3$ (**46**) from provitamin D$_3$ (**43**) was described by Fuse *et al.* and made use of a flow setup composed by two microflow quartz reactors equipped with a 400-W high-pressure Hg lamp with a Vycor filter. The procedure involved a first photochemical ring opening occurring in provitamin D_3 (**43**), followed by the thermal isomerization of the obtained previtamin D_3 (**44**). Unfortunately, due to the similar absorption spectra of reactant and photoproduct, a photochemical isomerization of the latter occurred, and the formation of side products, such as Tachisterol **45** was observed, limiting the overall yield of the process (not exceeding 20% on industrial scale).

In the developed device, two microreactors were connected with a PEEK tubing and the first reactor was irradiated with light in the 313–578 nm range. By contrast, the second reactor was irradiated at 360 nm light (the Hg lamp was here equipped with both a Vycor and a glass UV filter) and put on a hot oil bath at 100°C

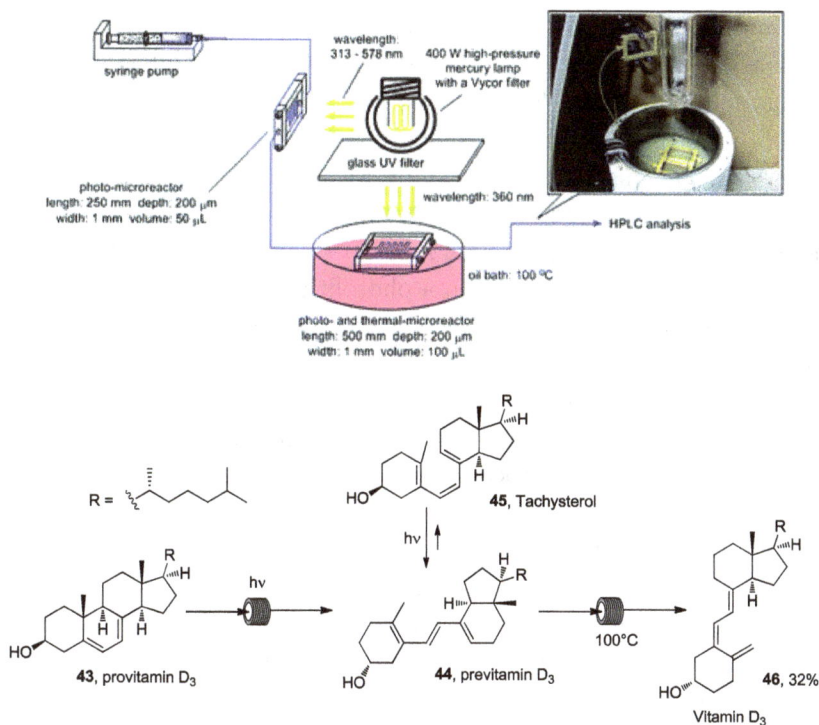

Fig. 1.29. Two-stage, continuous-flow synthesis of vitamin D_3 using two microreactors. (Adapted with permission from Ref. 81. Copyright © 2010 The Royal Society of Chemistry.)

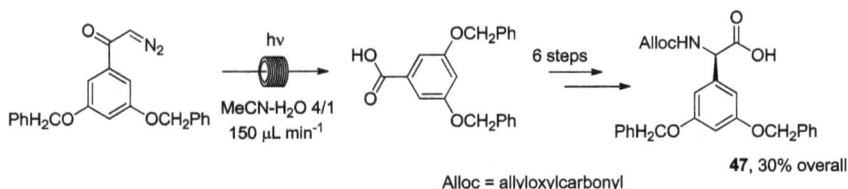

Scheme 1.3. Multistep synthesis of *N*-allyloxylcarbonyl 3,5-dihydroxyphenylglycine **47**.

(Figure 1.29).[81,83] Thermal isomerization of **44** took place in the second reactor, while continuous irradiation recovered the original equilibrium, thus favoring **44** at the expense of **45**. The new approach yielded desired **45** in 32% overall yield.

A microreactor flow apparatus was also exploited in the scalable multistep synthesis of *N*-allyloxylcarbonyl 3,5-dihydroxyphenylglycine **47**, a precursor of several bioactive compounds, including antibiotic vancomycin (Scheme 1.3).[83]

As hinted above, photochemical synthesis under microflow conditions can be carried out by having recourse to both commercially available and homemade devices. In particular, the wet etching technique allows for the home-made fabrication of chips. This procedure has been successfully applied on a photolithographic glass plate with a chromium layer covered by a photoresist. The latter was in turn removed in the areas exposed to UV light, according to the employed mask. The development of the chromium layer by etching made use of Chrometch hydrofluoric acid and allowed to obtain the desired depth channel. A silanization procedure finally allowed for the immobilization of porphyrins on the surface of the channels, and the obtained microreactor was used in photosensitized oxygenations (Figure 1.30).[84]

An improved singlet oxygen generation quantum yield was ensured by preventing dye oligomerization, often observed in homogeneous systems. Furthermore, the immobilization of the dye avoided the contamination of the target product and increased, at the same time, the turnover number of the process.[85]

A polyisobutylene-polymer-bound iridium(III) catalyst was used to promote deiodination of alkyl iodides or E/Z isomerization of

(a)

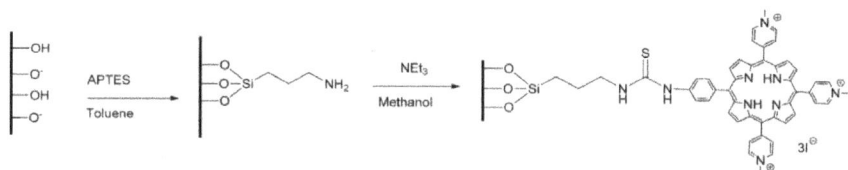

(b)

Fig. 1.30. (a) Chip design showing the 16 parallel channels. (b) Chip functionalized with a porphyrin. The photosensitizer was immobilized by the reaction of the isothiocyanate group with the amino group introduced by silanization with (3-aminopropyl)-triethoxysilane. (Reprinted with permission from Ref. 84. Copyright © 2012 American Chemical Society.)

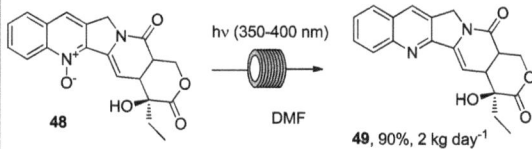

Fig. 1.31. Photochemical production unit for 10-hydroxycamptothecin **49**. (Reprinted with permission from Ref. 24. Creative Commons License.)

styrenes in a microreactor making use of a thermomorphic solvent system.[86] Here, a heptane/MeCN biphasic solution containing the photocatalyst (in heptane) and the reagents (in MeCN) was heated at 85°C to have a homogeneous mixture. Irradiation of this system with blue LED light ($\lambda = 455\,\text{nm}$) followed by cooling and phase separation allowed to isolate the products from the heptane phase containing the catalyst, that could be reused.[86]

Flow photoreactors were also sparsely used in industry. An industrial plant consisting of twelve microreactor units operated in parallel was used in the production of anticancer 10-hydroxycamptothecin **49** from the corresponding N-oxide **48** on a kg-scale. Each microreactor was made of two parallel quartz glass plates separated by a thin spacer, giving a 40–100-μm thin film of the reacting solution. High-pressure mercury lamps with spectral filters (350–400 nm) and an optical power output of 250 W were positioned on both sides of the microfilm reactor (Figure 1.31).[87]

1.7 Conclusion

As highlighted in this chapter, the assembling of a flow photoreactor is in most cases very easy and even novices can, in principle, build their own system according to their needs. The structure and the design of the photoreactor depend on the reaction studied and the light absorbing substrate.

Thus, through the simple UV–Vis spectrum measurement of such a substrate, it is possible to identify the most suitable light source and the material (that must be transparent and chemically inert) to be used in assembling the photoreactor. The power of the lamp is another key issue, since in some cases an unproductive decomposition of the substrate has been observed when using a too much powerful light source.

The scale of the process in turn determines the type of the device, being either a micro or mesophotoreactor. As for the latter class, the simplest and most versatile approach is to build a Booker-Milburn type reactor, in which irradiation takes place internally. This versatile system allows to use all of the light sources commercially available (medium- and low-pressure Hg lamps, CFL lamps, LED strips, etc.), thus making the optimization procedure easier. Furthermore, in this case the productivity can be tuned by changing the reactor channel dimensions. On the other hand, the small size of microflow reactor channels provides a way to integrate the light source on the microreactor assembly. Here, a scale-up is possible by connecting different microreactors in parallel, leaving untouched all the parameters of the device. For this apparatus, a punctual light is required and LEDs are the ideal (the only) source to be used, though the cost of LEDs with a $\lambda < 365\,$nm still represents a major stumbling block. Moreover, LEDs have a beam width often wider than the microreactor channels, resulting in an energy efficiency and productivity uncompetitive with classic immersion-well batch reactors.[22-28]

References

1. T. Wirth, *Microreactors in Organic Synthesis and Catalysis* (Wiley-VCH, Weinheim, 2008).
2. R. L. Hartman, J. P. McMullen and K. F. Jensen, *Angew. Chem. Int. Ed.* **50**, 7502 (2011).
3. B. P. Mason, K. E. Price, J. L. Steinbacher, A. R. Bogdan and D. T. McQuade, *Chem. Rev.* **107**, 2300 (2007).
4. K. Geyer, T. Gustafsson and P. H. Seeberger, *Synlett.* **15**, 2382 (2009).
5. A. Kirschning, W. Solodenko and K. Mennecke, *Chem.-Eur. J.* **12**, 5972 (2006).
6. C. Wiles and P. Watts, *Eur. J. Org. Chem.* **2008**, 1655 (2008).
7. S. V. Ley and I. R. Baxendale, *Proc. Bosen Symposium, Systems Chemistry* p. 65 (2008).
8. K. Jaehnisch, V. Hessel, H. Loewe and M. Baerns, *Angew. Chem. Int. Ed.* **43**, 406 (2004).
9. D. Webb and T. F. Jamison, *Chem. Sci.* **1**, 675 (2010).
10. D. Cambié, C. Bottecchia, N. J. W. Straathof, V. Hessel and T. Noël, *Chem. Rev.* **116**, 10276 (2016).
11. J. P. Knowles, L. D. Elliott and K. I. Booker-Milburn, *Beilstein J. Org. Chem.* **8**, 2025 (2012).
12. K. Gilmore and P. H. Seeberger, *Chem. Rec.* **14**, 410 (2014).
13. A. M. Braun, M. T. Maurette and E. Oliveros, *Photochemical Technology* (Wiley, Chichester, 1991).
14. I. Ninomiya and T. Naito, *Best Synthetic Methods: Photochemical Synthesis* (Academic Press, London, 1989).
15. C. Jung, K.-H. Funken and J. Ortner, *Photochem. Photobiol. Sci.* **4**, 409 (2005).
16. D. Dondi, S. Protti, A. Albini, S. Mañas Carpio and M. Fagnoni, *Green Chem.* **11**, 1653 (2009).
17. E. E. Coyle and M. Oelgemoeller, *Photochem. Photobiol. Sci.* **7**, 1313 (2008).
18. C. M. Doede and C. A. Walker, *Chem. Eng.* **62**, 159 (1955). For early efforts in the field of photochemical reaction engineering, see also: A. E. Cassano, P. L. Silveston and J. M. Smith, *Eng. Ind. Eng. Chem.* **59**, 18 (1967).
19. R. H. Feehs and N. J. Woodstown, U.S. Patent 3554887 (1971).
20. C. Birr, W. Lochinger, G. Stahnke and P. Lang, *Justus Liebigs Ann. Chem.* **763**, 162 (1972).

21. Y. Su, N. J. W. Straathof, V. Hessel and T. Noël, *Chem.-Eur. J.* **20**, 10562 (2014).
22. T. Noël and X. Wang and V. Hessel, *Chim. Oggi* **31**, 10 (2013).
23. M. Oelgemoeller, *Chem. Eng. Technol.* **35**, 1144 (2012).
24. M. Oelgemoeller and O. Shvydkiv, *Molecules* **16**, 7522 (2011).
25. Y. Matsushita, T. Ichimura, N. Ohba, S. Kumada, K. Sakeda, T. Suzuki, H. Tanibata and T. Murata, *Pure Appl. Chem.* **79**, 1959 (2007).
26. Z. J. Garlets, J. D. Nguyen and C. R. J. Stephenson, *Isr. J. Chem.* **54**, 351 (2014).
27. E. M. Schuster and P. Wipf, *Isr. J. Chem.* **54**, 361 (2014).
28. D. M. Roberge, B. Zimmermann, F. Rainone, M. Gottsponer, M. Eyholzer and N. Kockmann, *Org. Process Res. Dev.* **12**, 905 (2008).
29. G. E. Imoberdorf, F. Taghipour, M. Keshmiri and M. Mohseni, *Chem. Eng. Sci.* **63**, 4228 (2008).
30. A. Albini and M. Fagnoni, *Photochemically-Generated Intermediates in Synthesis* (Wiley, Hoboken, 2013).
31. S. Protti and M. Fagnoni, *Photochem. Photobiol. Sci.* **8**, 1499 (2009).
32. A. C. Gutierrez and T. F. Jamison, *J. Flow Chem.* **1**, 24 (2011).
33. C. Reichardt and T. Welton, *Solvents and Solvent Effects in Organic Chemistry*, 4th Edn. (Wiley-VCH, Weinheim, 2011).
34. B. D. A. Hook, W. Dohle, P. R. Hirst, M. Pickworth, M. B. Berry and K. I. Booker-Milburn, *J. Org. Chem.* **70**, 7558 (2005).
35. M. D. Lainchbury, M. I. Medley, P. M. Taylor, P. Hirst, W. Dohle and K. I. Booker-Milburn, *J. Org. Chem.* **73**, 6497 (2008).
36. T. P. Willumstad, O. Haze, X. Y. Mak, T. Y. Lam, Y.-P. Wang and R. L. Danheiser, *J. Org. Chem.* **78**, 11450 (2013).
37. D. Šterk, M. Jukič and Z. Časar, *Org. Process Res. Dev.* **17**, 145 (2013).
38. M. Hurevich, J. Kandasamy, B. M. Ponnappa, M. Collot, D. Kopetzki, D. T. McQuade and P. H. Seeberger, *Org. Lett.* **16**, 1794 (2014).
39. F. Wojcik, A. G. O'Brien, S. Goetze, P. H. Seeberger and L. Hartmann, *Chem.-Eur. J.* **19**, 3090 (2013).
40. M. Fagnoni, F. Bonassi, A. Palmieri, S. Protti, D. Ravelli and R. Ballini, *Adv. Synth. Catal.* **356**, 753 (2014).
41. F. Bonassi, D. Ravelli, S. Protti and M. Fagnoni, *Adv. Synth. Catal.* **357**, 3687 (2015).
42. M. Bergami, S. Protti, D. Ravelli and M. Fagnoni, *Adv. Synth. Catal.* **358**, 1164 (2016).
43. B. Shen, M. W. Bedore, A. Sniady and T. F. Jamison, *Chem. Commun.* **48**, 7444 (2012).
44. B. Shen and T. F. Jamison, *Aust. J. Chem.* **66**, 157 (2013).
45. K. G. Maskill, J. P. Knowles, L. D. Elliott, R. W. Alder and K. I. Booker-Milburn, *Angew. Chem. Int. Ed.* **52**, 1499 (2013).
46. Y. Zhang, M. L. Blackman, A. B. Leduc and T. F. Jamison, *Angew. Chem. Int. Ed.* **52**, 4251 (2013).

47. D. C. Harrowven, M. Mohamed, T. P. Gonçalves, R. J. Whitby, D. Bolien and H. F. Sneddon, *Angew. Chem. Int. Ed.* **51**, 4405 (2012).
48. S. Garbarino, L. Banfi, R. Riva and A. Basso, *J. Org. Chem.* **79**, 3615 (2014).
49. V. D. Pinho, B. Gutmann and C. O. Kappe, *RSC Adv.* **4**, 37419 (2014).
50. A. Yavorskyy, O. Shvydkiv, N. Hoffmann, K. Nolan and M. Oelgemoeller, *Org. Lett.* **14**, 4342 (2012).
51. D. Cantillo, O. de Frutos, J. A. Rincón, C. Mateos and C. O. Kappe, *J. Org. Chem.* **79**, 8486 (2014).
52. D. Cantillo, O. de Frutos, J. A. Rincón, C. Mateos and C. O. Kappe, *Org. Lett.* **16**, 896 (2014).
53. D. Cantillo, O. de Frutos, J. A. Rincon, C. Mateos and C. O. Kappe, *J. Org. Chem.* **79**, 223 (2014).
54. M. B. Nodwell, A. Bagai, S. D. Halperin, R. E. Martin, H. Knust and R. Britton, *Chem. Commun.* **51**, 11783 (2015).
55. K. Zeitler, *Angew. Chem. Int. Ed.* **48**, 9785 (2009).
56. D. M. Schultz and T. P. Yoon, *Science* **343**, 1239176 (2014).
57. J. Xuan and W.-J. Xiao, *Angew. Chem. Int. Ed.* **51**, 6828 (2012).
58. J. M. R. Narayanam and C. R. J. Stephenson, *Chem. Soc. Rev.* **40**, 102 (2011).
59. C. K. Prier, D. A. Rankic, D. W. C. MacMillan, *Chem. Rev.* **113**, 5322 (2013).
60. D. Ravelli, S. Protti and M. Fagnoni, *Chem. Rev.* **116**, 9850 (2016).
61. Z. He, M. Bae, J. Wu and T. F. Jamison, *Angew. Chem. Int. Ed.* **53**, 14451 (2014).
62. M. Neumann and K. Zeitler, *Org. Lett.* **14**, 2658 (2012).
63. R. S. Andrews, J. J. Becker and M. R. Gagné, *Angew. Chem. Int. Ed.* **51**, 4140 (2012).
64. M. Baumann and I. R. Baxendale, *React. Chem. Eng.* **1**, 147 (2016).
65. A. A. Assadi, A. Bouzaza and D. Wolbert, *J. Photochem. Photobiol. A: Chem.* **236**, 61 (2012).
66. S. Bachollet, K. Terao, S. Aida, Y. Nishiyama, K. Kakiuchi and M. Oelgemoeller, *Beilstein J. Org. Chem.* **9**, 2015 (2013).
67. F. R. Bou-Hamdanab and P. H. Seeberger, *Chem. Sci.* **3**, 1612 (2012).
68. J. W. Tucker, Y. Zhang, T. F. Jamison and C. R. J. Stephenson, *Angew. Chem. Int. Ed.* **51**, 4144 (2012).
69. J. D. Nguyen, B. Reiss, C. Dai and C. R. J. Stephenson, *Chem. Commun.* **49**, 4352 (2013).
70. J. D. Nguyen, E. M. D'Amato, J. M. R. Narayanam and C. R. J. Stephenson, *Nat. Chem.* **4**, 854 (2012).
71. N. J. W. Straathof, H. P. L. Gemoets, X. Wang, J. C. Schouten, V. Hessel and T. Noël, *ChemSusChem* **7**, 1612 (2014).
72. F. Levesque and P. H. Seeberger, *Angew. Chem. Int. Ed.* **51**, 1706 (2012).
73. D. Kopetzki, F. Lévesque and P. H. Seeberger, *Chem.-Eur. J.* **19**, 5450 (2013).

74. Z. Amara, J. F. B. Bellamy, R. Horvath, S. J. Miller, A. Beeby, A. Burgard, K. Rossen, M. Poliakoff and M. W. George, *Nat. Chem.* **7**, 489 (2015).
75. D. B. Ushakov, K. Gilmore, D. Kopetzki, D. T. McQuade and P. H. Seeberger, *Angew. Chem. Int. Ed.* **53**, 557 (2014).
76. M. Woźnica, N. Chaoui, S. Taabache and S. Blechert, *Chem.-Eur. J.* **20**, 14624 (2014).
77. C. Y. Park, Y. J. Kim, H. J. Lim, J. H. Park, M. J. Kim, S. W. Seo and C. P. Park, *RSC Adv.* **5**, 4233 (2015).
78. M. Nettekoven, B. Puellmann, R. E. Martin and D. Wechsler, *Tetrahedron Lett.* **53**, 1363 (2012).
79. I. G. T. M. Penders, Z. Amara, R. Horvath, K. Rossen, M. Poliakoff and M. W. George, *RSC Adv.* **5**, 6501 (2015).
80. L. M. Kreis, S. Krautwald, N. Pfeiffer, R. E. Martin and E. M. Carreira, *Org. Lett.* **15**, 1634 (2013).
81. S. Fuse, N. Tanabe, M. Yoshida, H. Yoshida, T. Doi and T. Takahashi, *Chem. Commun.* **46**, 8722 (2010).
82. S. Fuse, Y. Mifune, N. Tanabe and T. Takahashi, *Org. Biomol. Chem.* **10**, 5205 (2012).
83. Y. Mifune, S. Fuse and H. Tanaka, *J. Flow Chem.* **4**, 173 (2014).
84. E. K. Lumley, C. E. Dyer, N. Pamme and R. W. Boyle, *Org. Lett.* **14**, 5724 (2012).
85. J. Wahlen, D. E. De Vos, P. A. Jacobs and P. L. Alsters, *Adv. Synth. Catal.* **346**, 152 (2004).
86. D. Rackl, P. Kreitmeier and O. Reiser, *Green Chem.* **18**, 214 (2016).
87. S. Werner, R. Seliger, H. Rauter and F. Wissmann, European Patent 2065387 (2009).

Chapter 2

A General Introduction to Transport Phenomena in Continuous-Flow Microreactors for Photochemical Transformations

Yuanhai Su

Department of Chemical Engineering
School of Chemistry and Chemical Engineering
Shanghai Jiao Tong University, Shanghai, P. R. China
y.su@sjtu.edu.cn

2.1 Introduction

As compared with batch reactors, microreactors show a great deal of obvious advantages in photochemical transformations, including large specific surface area, fast mixing, enhanced heat and mass transfer rates, increased radiation homogeneity, continuous-flow operation, reduced safety hazards and the ease of increasing throughput by numbering-up, etc.[1-6] Therefore, microreactors have been widely applied in photochemical processes, providing remarkable reaction pathways to construct complex molecular structures in synthetic organic chemistry. While many chemists try to understand the chemistry principles behind photochemistry, the engineering aspects associated in these processes receive less attentions. In particular, transport phenomena such as photon transport, heat and mass transfer associated with the photochemical transformations in microreactors are not systematically described, making the difficulty on the judicious choices of proper reactor types, light sources, operational zones and so on.[7] In this chapter, we give an overview of the transport phenomena characteristics in microreactors applied for photochemical processes. Some criteria which can be used for design considerations are deduced, and specific examples are given in order to further understand underlying engineering principles in photochemical processes conducted in microreactors. Moreover, this chapter aims to aid the reader on the recognition of the intrinsic characteristics of photomicroreactors and thus exploit their full potential in photochemical transformations.

2.2 Radiation Transport Phenomena

Obviously different from thermochemical processes, photochemical processes involve the transport of photons that provide a means to overcome energy barrier during chemical transformations.[8-11] The acceptors of photons can be reagents or photocatalysts, leading to two different excitation ways including direct and indirect excitations. Photochemical transformations provide opportunities to carry out a large number of reactions which are otherwise difficult to realize through classical thermochemical approaches. They are widely used in the fields of organic synthesis, material science, and environmental treatment.[8,9,12-14] Most photochemical processes are conducted under mild reaction conditions (e.g. room temperature and atmospheric pressure) and the use of low energy visible light as an abundant energy source is frequently encountered, making photochemistry to be known as a green synthetic method. Nevertheless, some limitations are typically associated with batch reactors when applying them in photochemical processes, which hinder extensive implementation of this energy transformation mode.[15-17] For instance, a limitation arises from the attenuation effect of photon transport which prevents classical scale-up of batch reactors via a dimension enlarging strategy.[18] Furthermore, the mixing efficiency, heat and mass transfer rates in batch reactors usually do not match intrinsic reaction rates especially for fast photochemical reactions. These limitations may be overcome by utilizing continuous-flow photomicroreactors.[7,19-24]

Photon absorption or light scattering leads to an inhomogeneous distribution of light intensity and different local reaction rates over the reactor. Thus, the overall conversion and the yield are highly depending on the radiation distribution within the reactor. The spectral specific intensity (I) is used to represent the radiation distribution in reaction medium, which constitutes the amount of irradiative energy streaming through a unit area perpendicular to the direction of propagation $(\underline{\Omega})$, per unit wavelength and per unit time[25]:

$$I(s, \underline{\Omega}, t, \lambda) = \lim_{dA, d\Omega, dt, d\lambda} \left(\frac{dE_\lambda}{dA \cos\theta d\underline{\Omega} dt d\lambda} \right) \tag{1}$$

where dE_λ is the total amount of irradiative energy passing through a surface in the time dt within a wavelength range between λ and $\lambda + d\lambda$. Based on the spectral specific intensity distribution, we can define the incident radiation (G_λ) and the local volumetric rate of energy absorption (LVREA) as follows:

$$G_\lambda(s, t) = \int_{\underline{\Omega}} I_{\lambda, \underline{\Omega}}(s, t) d\underline{\Omega} \tag{2}$$

$$\text{LVREA}_\lambda(s, t) = \kappa_\lambda G_\lambda(s, t) \tag{3}$$

where κ_λ is the volumetric absorption coefficient. It represents the fraction of the incident radiation that is absorbed by the matter per unit length along the path of the beam. The LVREA depends on the lamp emission, the geometry of the reactor and photophysical properties of the reaction mixture. It is difficult to obtain uniform distribution of LVREA over the reactor due to the attenuation effect. Encouragingly, this non-uniformity can be minimized in microreactors due to the small characteristic dimensions.[26,27]

The illumination efficiency can be represented by the photonic efficiency (ξ):

$$\xi = \frac{\text{rate of the reaction}}{\text{photon flux}} \tag{4}$$

$$\text{photon flux } (q) = \frac{G_\lambda \times \lambda}{hc} \tag{5}$$

where h is the Planck's constant, and c is the speed of light in vacuum, respectively. The photonic efficiency in microreactors (e.g. $\xi = 0.0262$) is about one order of magnitude higher than those in batch reactors (e.g. $\xi = 0.0086 - 0.0042$).[28-30] This value can be further improved by using microscale light sources (e.g. LEDs), which matches the dimensions of microchannels, or the optimization of light propagation with the help of optical fibers or mirrors as light reflectors.[31] Through the compact integration of a capillary microreactor with LED light source, the photonic efficiency could reach 0.66, which was about two orders of magnitude higher than batch photoreactors.[26] Figure 2.1 shows the combination of microreactors with frequently used light sources for photochemical transformations.

In a simplified reaction scheme A → B, the reaction rate can be expressed as follows[36]:

$$r = \phi_\lambda (\text{LVREA})_\lambda \qquad (6)$$

From this equation, it is easy to understand that effective energy input of the light sources affect LVREA, photon flux, and thus the reaction rate. This was convincingly demonstrated via the photocatalytic aerobic oxidation of thiols to disulfides in a photomicroreactor with the use of an AC/DC converter.[26] Higher power input resulted in higher LVREA or photon flux, which can subsequently excite the photocatalyst and trigger the photocatalytic transformations leading to the desired disulfide compounds. Figure 2.2 shows that the relationship between the yield and the effective energy input was approximately linear at various residence times.[26] This figure clearly indicates that the reaction rates in microreactors for photocatalytic transformations can be easily tailored by changing the power input or the photon flux.

Direct measurement on the spectral specific intensity in photomicroreactors with photometers (radiometers) has not been realized up to now. This is because the dimensions of microreactors are smaller than those of the sensors and also, in most cases, the irradiated area remains difficult to determine accurately.[7] Therefore, the photon flux in photomicroreactors cannot be calculated through the

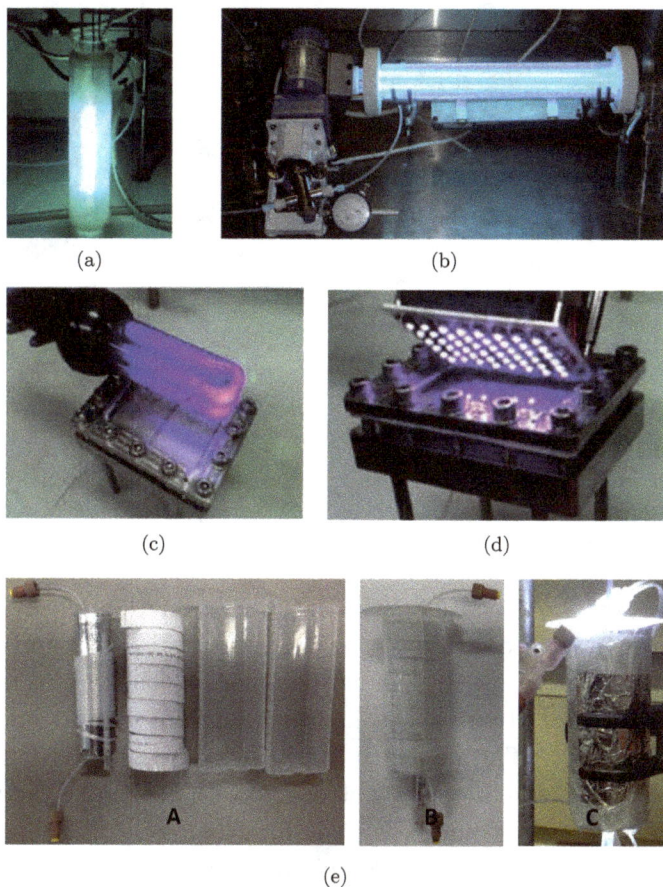

Fig. 2.1. UV photomicroreactors: (a) FEP continuous-flow reactor coiled around a 400-W medium-pressure Hg lamp. (Reprinted with permission from Ref. 32. Copyright © 2005 American Chemical Society.) (b) FEP continuous-flow reactor with a 36-W PL-L lamp (254 nm)[33] (Reprinted with permission from Ref. 33. Copyright © 2013 WILEY-VCH Verlag GmbH & Co. KGaA, Weinheim.) (c) A photomicroreactor with quartz top plate irradiated with a black light lamp (15 W, 360 nm). (Reprinted with permission from Ref. 34. Copyright © 2011 Akademiai Kiado.) (d) A photomicroreactor with quartz top plate irradiated with UV LEDs (1.7 W, 360 nm). Note that the light is better directed toward the microreactor. (Reprinted with permission from Ref. 34. Copyright © 2011 Akademiai Kiado.) (e) Photomicroreactor with LEDs as light source: A: different components of the photomicroreactor assembly, with (from left to right) the capillary microreactor coiled around an aluminum coated syringe, the LED stripe, and the outer syringe in which the photomicroreactor and LEDs are placed; B: assembled version; C: photomicroreactor assembly in operation. (Reprinted with permission from Ref. 35. Copyright © 2015 John Wiley & Sons Inc.)

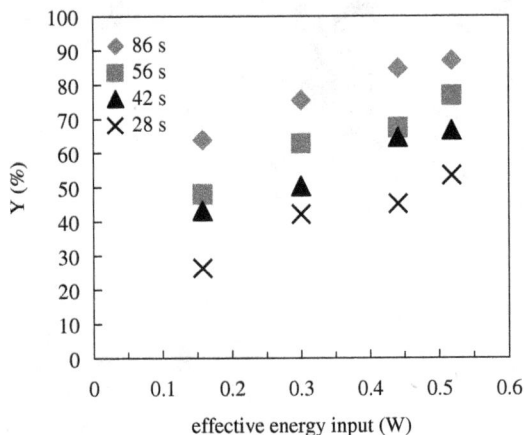

Fig. 2.2. Relationship between the yield and the effective energy input. (Reprinted with permission from Ref. 26. Copyright © 2015 WILEY-VCH Verlag GmbH & Co. KGaA, Weinheim.)

integration of the spectral specific intensity over the different directions of radiation propagation according to equations (2) and (3). An efficient alternative is to use an actinometer, which involves a simple photochemical reaction with a known quantum yield at quantified wavelengths.[37–40] The property of temperature independency further makes an actinometer as a strong tool to determine the photon flux in photoreactors. Loubiere *et al.* have reported on the measurement of photon flux in two photomicroreactors (A and B) and in an immersion-well batch photoreactor with the actinometry of potassium ferrioxalate.[41] Photomicroreactor A was constructed by winding fluorinated ethylene propylene (FEP) tubing with 508 μm i.d. and 4 m length, which was coiled around a Pyrex immersion well with an immersed high-pressure Hg lamp. Photomicroreactor B consisted of an FEP tube (508 μm i.d. and 2.65 m length) which was fixed in a channel carved in a flat aluminum plate, with illumination from UV-LED array. The batch photoreactor had the same immersion well and the same light source as those used for the photomicroreactor A. Their results indicated the significant difference between the photon flux actually received and the radiant power emitted by the light source and the importance of *in situ* measurements on the photon

Table 2.1. Photon flux (q_p) received, absorbed photon flux density (q_p/V_r) and photon flux density received at the reactor wall (L_p^w) at a wave length of 365 nm in the different photoreactors.[41]

Parameters	Photomicroreactor		Batch photoreactor
	A	B	
q_p (10^{-6} Einstein s^{-1})	4.07	0.382	7.40
q_p/V_r (10^{-6} Einstein s^{-1})	5.02	0.71	0.033
L_p^w (10^{-3} Einstein m^{-2} s^{-1})	2.55	0.36	0.17

flux. In other words, the reaction medium cannot ensure a complete absorption of photons emitted by the light source due to various factors such as emission spectra, composition of reaction medium, reactor exposition, reflectance and transmittance of the reactor materials, etc. Furthermore, the authors compared the received photon flux and the absorbed photon flux density in these two photomicroreactors and the batch photoreactor (see Table 2.1). The photon flux in the microreactor A was 4.07×10^{-6} Einstein s^{-1}, which was in the same order of magnitude but lower than the batch photoreactor (7.4×10^{-6} Einstein s^{-1}) when applying the same light source (cf., as polychromatic mercury lamp). However, the photon flux density in the photomicroreactor could reach 5.02 Einstein m^{-3} s^{-1}, which was much higher than that in the batch photoreactor (0.033 Einstein m^{-3} s^{-1}). This directly explains the reason why the times required to reach a given conversion in photomicroreactors are much shorter in comparison with those found in conventional batch photoreactors.

Based on the known photon flux in photoreactors, one can calculate the quantum yield (Φ) as follows:

$$\Phi = \frac{\text{number of product molecules formed}}{\text{number of photons absorbed by reactive medium}} \quad (7)$$

$$\text{Number of photons} = \frac{G_\lambda \lambda S t}{hc} = qSt \quad (8)$$

where S is the illumination area and t is the reaction time. The quantum yield can give some information on the reaction mechanisms of the photochemical processes. As a rule of thumb, it is assumed that when the quantum yield is lower than 1, a catalytic mechanism is valid and the reaction occurring is considered as a photocatalytic process. It is however possible that the quantum yield is much larger than 1. In this case, one photon initiates conversion of more than one reactant molecules. This type of reactions is defined as a photon induced radical chain reaction such as photochemical polymerizations. For this type of reactions, determination of the quantum yield can give insight in the average radical chain length.[42]

The full description of the radiation transport is rather complex since it is related to many factors, e.g. the light source, the geometry of photoreactor, the composition of reaction medium, and so on. For monochromatic light, the radiation transport equation (RTE) can be described by[25]:

$$\frac{dI_{\lambda,\,\underline{\Omega}}(s,\,t)}{ds} + \underbrace{k_\lambda(s,\,t)I_{\lambda,\,\underline{\Omega}}(s,\,t)}_{\text{Absorption}} + \underbrace{\sigma_\lambda(s,\,t)I_{\lambda,\,\underline{\Omega}}(s,\,t)}_{\text{out-scattering}}$$

$$= \underbrace{j_\lambda^e(s,\,t)}_{\text{Emission}} + \frac{\sigma_{\lambda,\,\underline{\Omega}}(s,\,t)}{4\pi}$$

$$\times \underbrace{\int_{\Omega'=4\pi} p(\underline{\Omega'} \to \underline{\Omega})\,I_{\lambda,\,\underline{\Omega'}}\,(s,\,t)d\underline{\Omega'}}_{\text{in-scattering}} \tag{9}$$

Relevant phenomena including absorption, emission and scattering effects are described in this equation, which can be encountered in photochemical processes. The solution of this equation provides the spectral specific intensity with regard to the distance along the direction of radiation propagation. The emission in reaction medium can be neglected when the photochemical reactions are carried out at low or medium temperatures. For homogeneous reaction mixtures, the scattering effects are rather weak and can therefore be neglected. If the radiation intensity from the light source is constant and the light propagation occurs along a single direction, the RTE for homogeneous photochemical reactions can be simplified as follows:

$$\frac{dI_{\lambda,\,\Omega}(s)}{ds} = -k_\lambda(s)I_{\lambda,\,\Omega}(s) \tag{10}$$

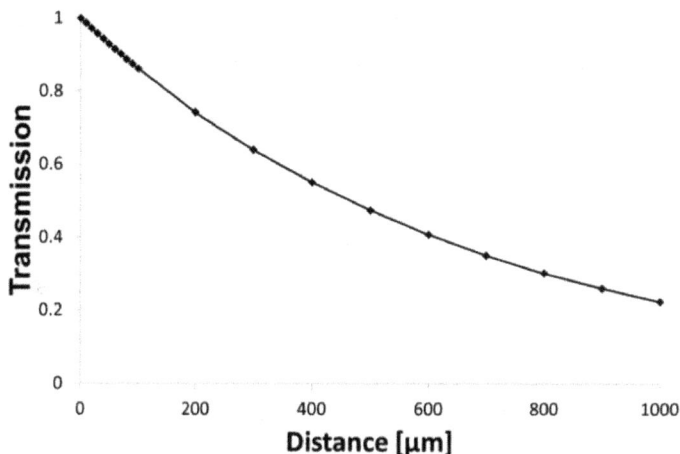

Fig. 2.3. Absorbance of incident light as a function of distance in the reaction medium containing $Ru(bpy)_3^{2+}$ as a photocatalyst. The profile is obtained by utilizing the Bouguer–Lambert–Beer law: concentration $c = 0.5$ mM; molar extinction coefficient $\varepsilon[Ru(bpy)_3^{2+}] = 13000\,cm^{-1}\,M^{-1}$; path length l; transmission T; spectral specific intensity in reaction medium closed to reactor wall I_0; spectral specific intensity after transmission I. (Reprinted with permission from Ref. 7. Copyright © 2014 WILEY-VCH Verlag GmbH & Co. KGaA, Weinheim.)

This equation can be further transformed to describe the light absorbance (A) in a reaction medium:

$$A = \log_{10} T = \log_{10} \frac{I_0}{I} = \varepsilon c l \qquad (11)$$

This is the well-known Bouguer–Lambert–Beer equation, indicating that the absorption is dependent on the molar extinction coefficient (ε), the concentration of the absorbing species (c), and the path length (l). The effects of characteristic dimensions of photoreactors on the light absorption can be easily illustrated for a given photocatalytic example using $Ru(bpy)_3Cl_2$ as a photocatalyst (see Figure 2.3).[19] $Ru(bpy)_3Cl_2$, due to its high molar extinction coefficient, absorbs a significant amount of the incident light in the first few hundreds of micrometers. 50% of the light irradiation is already absorbed within a distance of only $500\,\mu$m. This figure demonstrates that small dimensions of microreactors and thus high surface-to-volume ratios are crucial to achieve/approach an almost complete illumination homogeneity. In fact, some types of microreactors, such as interdigital micromixers with channel characteristic size of about 25 μm, show great advantages on achieving illumination homogeneity for photochemical transformations.[43–46] Given the excellent photon flux intensity through the whole microreactors, it is evident that shorter reaction times are often required compared to large-scale vessels. In addition, similar productivities can be obtained in microreactors while utilizing much lower photocatalyst amounts as in batch reactors.

2.3 Mixing Time and Criteria for Single-Phase Reaction Systems

Small immersion-well photoreactors in conjunction with mercury-vapor discharge lamps are widely applied in photochemical transformations in research laboratories, which are categorized as a kind of batch or semi-batch reactors.[47–49] However, transport properties such as mixing, mass and heat transfer in large batch reactors are difficult to maintain the same values as those in small batch reactors. New types of reactors such as spinning disc reactors, monolithic reactors, thin film reactors and microreactors have been developed in order to overcome mass transport limitations in photochemical transformations.[50] In particular, microreactors can provide even higher surface-to-volume ratios and mass transport rate than other novel reactors due to their smaller characteristic dimensions. For single-phase systems, the characteristic mixing time is typically used to describe the mass transport phenomena. In microchannels, most operations are conducted at low Reynolds numbers ($Re < 100$) and a strict laminar flow is observed due to the small length scales. In this case, the mixing is driven by molecular diffusion and the characteristic mixing time (t_m) in microchannels strongly depends on the diffusion length (L) and the molecular diffusivity (D) according to the Einstein–Smoluchovski equation:

$$t_m = \frac{L^2}{D} \tag{12}$$

This principle shows that smaller dimensions will result in faster mixing. The development of passive micromixers such as interdigital micromixer, split-and-recombine micromixer, split-and-recombine micromixers, packed-bed microchannels, caterpillar micromixers and zigzag microchannels is remarkable, in which the diffusion distance is reduced and the convection is introduced through decreasing the characteristic dimensions or fabricating special structures.[51] If the convection and the chaotic advection are taken into account on the improvement of the mixing efficiency, the characteristic mixing time is calculated as follows[52]:

$$t_m = \frac{24.5L^2}{U}(\mu Re^3)^{-1/2} \tag{13}$$

The selection of suitable reactors for reaction processes is highly dependent on the characteristics of the intrinsic reaction kinetics. The characteristic reaction time (t_r) is used to judge whether the problem needs to be solved as an equilibrium problem or a kinetic problem. It also can provide information on whether the effect of mixing on the reaction performance exists in a specific reaction process through the comparison with the characteristic mixing time. For thermochemical transformations, the characteristic reaction time is dependent on the reaction rate constant and the reactant concentration. In thermochemical processes, the reaction rate constant can be easily varied by changing the reaction temperature according to the Arrhenius equation. In contrast, the intrinsic reaction rate (r) for photochemical transformations is strongly related to the LVREA and the quantum yield (Φ), as shown in equation (14). The characteristic reaction time can be defined as the following equation[53]:

$$t_r = \frac{c_0}{r} = \frac{c_0}{\phi_\lambda(\text{LVREA})_\lambda} \tag{14}$$

where c_0 is the initial reactant concentration. Therefore, the characteristic reaction time is inversely proportional to both the LVREA and the quantum yield (Φ) in the reaction scheme (e.g. A $\rightarrow$ B). With both the characteristic mixing time and the characteristic reaction time in hand, one can calculate the second Damköhler number

(DaII), which is used to describe the relative importance of mass transport effects on reaction processes[4]:

$$Da\text{II} = \frac{t_m}{t_r} \qquad (15)$$

The effect of mixing on the reaction performance will be eliminated if DaII is smaller than 1. Otherwise, a concentration gradient of the compounds will exist in the reaction medium which might result in the generation of an extensive amount of by-products. According to the value of DaII, the reaction regime can be classified as three types:

DaII < 1, reaction rate limited regime;
DaII ≈ 1, mixed mass transport-reaction rate limited regime;
DaII > 1, mass transport controlled regime.

The calculation of DaII for a given chemical reaction will confirm whether mixing effects occur and, consequently, if the use of more advanced continuous-flow mixers or reactors is required. In addition, the residence time should be longer than the characteristic reaction time in order to reach full conversion, which can be expressed by the first Damköhler number (DaI):

$$Da\text{I} = \frac{t_p}{t_r} \gg 1 \qquad (16)$$

In microreactors or microchannels, the residence time can be easily varied by controlling the flow rates or the lengths of the channels. An increase in the flow rate usually results in shorter residence time and higher mixing efficiency which indicates shorter characteristic mixing time.

In photochemical processes, the photon transport and the mass transport interact with each other, which complicates the design of a photoreactor and the process modeling.[25,54–57] The concentration distribution of reagents or photocatalysts consisting of chromophores affects the absorption and distribution of photons in photoreactors. Meanwhile, the different local photochemical reaction rates over the reactor result in concentration gradients. A heterogeneous velocity field along the transverse direction relative to the flow direction further exaggerates concentration gradients. For example, in the single-phase systems, the parabolic velocity profile exists in a photomicroreactor, and the molecules close to the walls are transported to the places with lower concentrations by diffusion or convection. The concentration gradients generated will annihilate if the transverse mixing time is smaller than the residence time, that is, if the Fourier number *Fo* is higher than 1 ($Fo \geq 1$).[58] On the other hand, the light attenuation along the optical light path also can result in the concentration gradients, which is distinctly different from the thermochemical processes. Depending on the emission characteristics of light sources and the absorption properties of reaction mixture, the reaction rate may be heterogeneous in the transverse direction and will, thus, contribute to generate some concentration gradients. These gradients will induce a mass flux by molecular diffusion, which tends to oppose to the gradients generated. The relationship between DaI and DaII can be linked by the following equation:

$$Fo = \frac{Da\text{I}}{Da\text{II}} \tag{17}$$

For a given set of these three dimensionless numbers (DaI, DaII, *Fo*), one can identify whether concentration gradients will exist along the transverse direction at the outlet of the photomicroreactor due to the effect of heterogeneous velocity field or the heterogeneous kinetic rate field (light attenuation).[58] Figure 2.4 shows the relationship among these three dimensionless numbers, which can be applied for choosing an appropriate operational zone in a specific photochemical process.

Four operational zones can be distinguished in Figure 2.4. In Zone D, both DaII and $1/Fo$ are lower than 1, indicating that there are no concentration gradients in the transverse section and there are no diffusion limitations in photochemical processes. This is because the transverse diffusion time is smaller than both the residence time and the reaction time. In the subdivided zone (DaI < 1), the reaction is not completely finished in the microreactor and the conversion

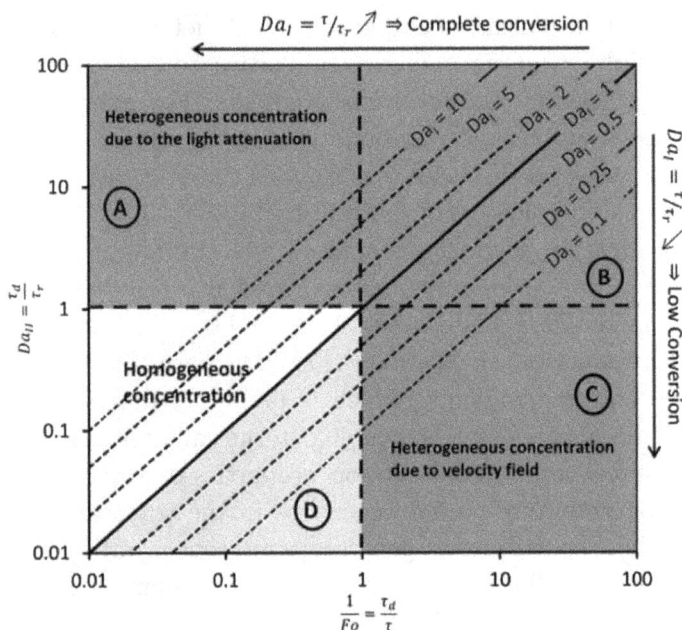

Fig. 2.4. Different operational zones for a specific photochemical process based on the relationship among $Da\mathrm{I}$, $Da\mathrm{II}$ and Fo. (Reprinted with permission from Ref. 58. Copyright © 2015 John Wiley & Sons Inc.)

changes with the variation of the residence time. One can use the experimental data obtained in this subzone for studying the reaction kinetics of a photochemical process or screening operational parameters for process optimization.[27] In another subdivided zone ($Da\mathrm{I} \geq 1$), the conversion is close to 1 at the outlet of the microreactor, and one can conduct the photochemical transformations for production processes.

In Zone A ($Da\mathrm{I} \geq 1$ and $1/Fo \leq 1$), the heterogeneous kinetic rate field results in the concentration gradients.[58] This zone is not ideal for the use of the microreactor since the reaction product is over-exposed to the radiation, possibly leading to the occurrence of side reactions (e.g. photodecomposition). Even the conversion is complete before the outlet of the microreactor, the operations in this zone are not recommended. In Zone C ($Da\mathrm{I} \leq 1$ and $1/Fo \geq 1$), the concentration gradients exist because of the heterogeneous velocity field, so that the intrinsic kinetic data cannot be obtained. Zone B corresponds to $DaI \geq 1$ and $1/Fo \geq 1$, in which the concentration gradients are generated due to both the heterogeneous velocity and kinetic rate fields. The conversion or the selectivity will be significantly affected by the occurrence of these concentration gradients, and the reactor performance is far away from a plug flow behavior in this zone (Zone B). It is worth noting that small dimensions of photomicroreactors induce the decease of the values of $DaII$ and $1/Fo$, making the microreactors typically operating in Zone D.[26,59]

The effect of mixing efficiency on the photochemical transformations was well demonstrated by Oelgemoeller *et al.*[60] They investigated the DMBP-sensitized addition of isopropanol to furanones in a batch reactor and in three different microreactor systems, including a microdwell device (0.5×2 mm ($D \times W$)), a microchip (0.15×0.15 mm ($D \times W$)) and a capillary microreactor (0.56 mm i.d.). It was observed that the continuous-flow microreactor systems provided higher or comparable chemical conversions than the batch reactor. The LED-driven microchip gave the best overall results with regard to conversion rate, reactor geometry and energy efficiency. This was attributed to the higher illumination efficiency (LVREA) which can be achieved when using small light sources (UV-LEDs) and the higher mixing efficiency (shorter characteristic mixing time) arising from the confined spaces in the microchip system. Meanwhile, the characteristic mixing time, the characteristic reaction time and the residence time matched each other in the microchip. In addition, the capillary photomicroreactor systems are considered as modular, inexpensive, effective and operationally simple designs, which can be prepared rapidly from commercially available parts within 1 h even by non-specialists.[35,61]

2.4 Mass Transfer and Criteria for Multiphase Reaction Systems

Multiphase reactions at least involve two phases, in which the reactants transfer from one phase to the other phase and react with other reactants. Typical multiphase reaction systems include gas–liquid, liquid–liquid, liquid–solid, and their combinations. The mass transfer of reactants might be the rate determining step, depending on the intrinsic reaction kinetics and the transport properties of reactors.[62-65] The importance of fast mass transfer in multiphase photochemical reactions can be demonstrated by the direct oxygenation of organic molecules with singlet oxygen as an oxidant.[66-68] In these gas–liquid biphasic photochemical processes, singlet oxygen is typically prepared via a dye-sensitized photoexcitation of triplet oxygen and can be used as an efficient oxidant in a variety of interesting synthetic reactions.[69] First, oxygen in the gas phase is transferred to the organic phase and then transformed to singlet oxygen under the effect of a photocatalyst (sensitizer) illuminated by the light. Singlet oxygen subsequently reacts with organic molecules to form e.g. peroxide intermediates. In such processes, the interfacial mass transfer of oxygen affects the distribution of sensitized singlet oxygen in the reaction medium and thus both yield and selectivity.

Furthermore, the mass transfer characteristics in reactors should be known in order to carry out reactor design and process optimization. The overall volumetric mass transfer coefficient $(k_L a)$ is usually used to describe the interfacial mass transfer rates. The values of $k_L a$ in microreactors are at least one to two orders of magnitude higher than those obtained in conventional reactors or contactors.[1,70-74] Some correlations have been developed to predict the gas–liquid mass transfer characteristics in microreactors or microchannels, according to different flow patterns. That is, for segmented flow (so-called Taylor flow or slug flow)[75]:

$$Sh_L a d_h = 0.084 Re_G^{0.213} Re_L^{0.937} Sc_L^{0.5} \qquad (18)$$

For slug-annular flow and churn flow:

$$Sh_L a d_h = 0.058 Re_G^{0.344} Re_L^{0.912} Sc_L^{0.5} \qquad (19)$$

where Sh_L is the liquid Sherwood number defined by $k_L d_h / D$, Re_G is the superficial gas Reynolds number defined by $\rho_G d_h j_G / \mu_G$, Re_L is the superficial liquid Reynolds number defined by $\rho_L d_h j_L / \mu_L$, Sc_L

Table 2.2. Comparison of mass transfer parameters and effective interfacial area for microreactors and conventional reactors.

Type of reactor/contactor	$k_L a \times 10^2$ (s^{-1})	$k_L \times 10^2$ (m s^{-1})	a (m^2/m^3)
Bubble column	0.5–24	10–40	50–600
Couette–Taylor flow reactor	3–21	9–20	200–1200
Impinging jet absorber	2.5–122	29–66	90–2050
Packed column	0.04–102	4–60	10–1700
Spray column	1.5–2.2	12–19	75–170
Static mixer	10–250	100–450	100–1000
Stirred tank	3–40	0.3–80	100–2000
Tube reactor	0.5–70	10–100	50–700
Microreactor	30–2100	40–160	3400–9000

is the liquid Schmidt number defined by $\mu_L / \rho_L D$, k_L is the liquid side mass transfer coefficient, d_h is the hydraulic diameter of microchannels, j_G and j_L are the superficial gas and liquid velocities, ρ_G and ρ_L are the densities of gas and liquid phases and μ_G and μ_L are the superficial viscosities of gas and liquid phases, respectively.

Compared to conventional reactors, the substantial increase in ka in microreactors mainly originates from the large area-to-volume ratios or effective interfacial area (a). The small characteristic dimensions of microreactors can result in very high effective interfacial area up to $9000 \, \text{m}^2/\text{m}^3$, which is rather difficult to achieve in conventional devices. Table 2.2 gives an overview of the $k_L a$ values for a variety of multiphase reactors.[76–80]

It should be noted that the hydrodynamics of multiphase systems in microreactors strongly depends on the flow rates.[81] A higher flow rate leads to a better dispersion of gas–liquid or liquid–liquid two-phases in the microreactors. Consequently, the effective interfacial area and the mass transfer rate increase with increasing flow rates in the microreactors. This is especially useful for achieving fast reactions with short-living species in multiphase photochemical transformations, such as the aforementioned gas-liquid photochemical processes involving singlet oxygen.[82] The lifetime of singlet oxygen is extremely short (e.g. $9.7 \, \mu s$ in ethanol) and thus fast mass transfer is required to make efficient use of the generated active species.[83] Moreover, the high reactivity of singlet oxygen can result into the formation of many by-products upon prolonged irradiation times. Therefore, such kinds of gas–liquid photoreactions are difficult to successfully carry out in large-scale batch reactors with high selectivity and yield due to low effective interfacial area and mass transfer rate. In contrast, microreactors can provide higher mass transfer rates and better control on reaction times, showing significant application potential on the transformations of short-living species in photochemical processes.[7]

If the mass transfer rate of reactive compounds is fast enough, the reaction will occur inside the bulk phase in multiphase processes. Otherwise, the mass transfer rate cannot match the intrinsic reaction rate, and thus the reaction will mainly take place in the interface between different phases. Different reaction regimes including slow, fast and instantaneous regimes can be distinguished according to the value of a dimensionless parameter (i.e. Hatta number). For gas–liquid and liquid–liquid biphasic reaction systems, the Hatta number is used to compare the reaction rate in a reactive film and the diffusion rate through this film.[53,84,85] For a reaction with mth order of A and nth order of B, where the reactant A transfers to the bulk phase and reacts with the reactant B, Ha can be described as the following equation:

$$Ha = \frac{\sqrt{2/(m+1)k_{m,\,n}(c_{A,\,i})^{m-1}(c_{B,\,\text{bulk}})^{n}D_{A}}}{k_{L}} \qquad (20)$$

where m and n are the reaction orders for gaseous reactants and substrates, $C_{A,i}$ is the interfacial concentration of reactant A between two phases, and $C_{B,\,\text{bulk}}$ is the concentration of reactant B in the bulk phase, respectively. When $Ha < 0.3$, the reaction is in the slow reaction regime and the reaction takes place in the bulk phase without mass transfer limitations. When $Ha > 3$, the reaction is in the instantaneous regime and the reaction occurs at the gas–liquid interface. In this case, the reaction process is controlled by the mass transfer rate. When $0.3 < Ha < 3$, the reaction is in the transitional regime and it is dominated by both the mass transfer rate and the intrinsic reaction rate. Figure 2.5 shows a schematic diagram of a gas–liquid

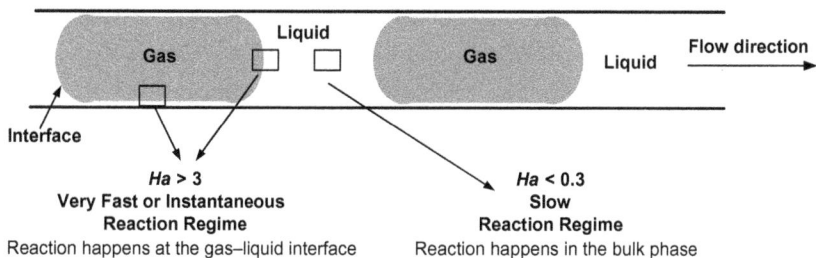

Fig. 2.5. Schematic representation of the different reaction regimes according to the Hatta number in a segmented gas–liquid photochemical transformation. (Reprinted with permission from Ref. 7. Copyright © 2014 WILEY-VCH Verlag GmbH & Co. KGaA, Weinheim.)

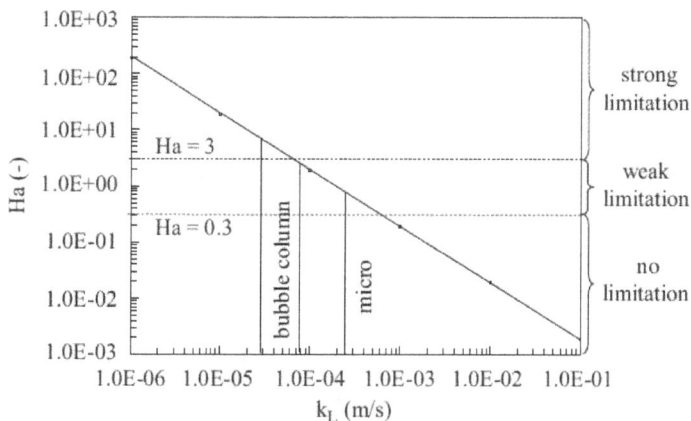

Fig. 2.6. Hatta number versus mass transfer coefficient with $k_r = 9.49 \times 10^{-1}$/s, $D = 3.08 \times 10^{-8}$ m^2/s, k_L for bubble columns in range of 3.7×10^{-5}–8.2×10^{-5} m/s, and k_L for capillary microreactors larger than 2.5×10^{-4}. (Reprinted with permission from Ref. 86. Copyright © 2010 Elsevier Ltd.)

two-phase segmented flow in microchannels, which is widely used in organic photochemistry. As shown in Figure 2.5, the reaction regimes of gas–liquid reactions are classified into different regimes depending on the value of Hatta number.

The dependence of the Hatta number on the mass transfer coefficient for the uncatalyzed selective aerobic oxidation of liquid cyclohexane directly indicates the advantages of microreactors as compared with conventional multiphase reactors.[86] As shown in

Figure 2.6, there are no mass transfer limitations when conducting this reaction in a capillary microreactor ($Ha < 0.3$). This means that the reaction process occurs inside the bulk phase due to the fast supply of oxidant from the gas to the liquid phase. However, mass transfer limitations still occur when bubble column reactors are used. Recently, Noël *et al.* have calculated Ha for the photocatalytic aerobic oxidation of thiols to disulfides and found that the value was even lower than 0.06.[35] This low value indicates that the reaction occurs in the bulk phase and that no mass transfer limitations are present in this gas–liquid photocatalytic process. The absence of mass transfer limitations allows one to determine intrinsic kinetics with microreactor technology.

2.5 Energy Transformation and Heat Transfer

In photochemical processes, upon light absorption a molecular system is transferred from the ground state to an electronically excited state, which facilitates chemical transformations that are otherwise difficult to reach by thermochemical or electrochemical activation. Electrical energy, optical energy and chemical energy are in reciprocal transformations in photochemical processes.[87–89] Actually, the energy transformation and heat transfer phenomena in photochemical processes are rather complex. Heat in photoreaction systems can be produced by both the light source and the reaction process. The heat produced by the light source is highly related to the photoelectric transformation efficiency. In order to improve the energy utilization efficiency in photochemical processes, one should select proper light sources with their emission spectra matching the absorption characteristics of reaction medium. Moreover, it is generally known that the efficiency of LED light sources that are widely applied in continuous-flow photochemical transformations decreases with increasing temperatures (so-called temperature efficiency droop).[90] Therefore, the heat management should be taken into account when designing a photochemical system.[91]

In the most general cases, the temperature (T) in continuous-flow reactors is a function of both the spatial coordinates and other operational conditions such as flow rates of the feed. With fixed illumination from the light source and flow rates, the temperature profile of reaction medium in photochemical processes depends on the optical absorption of reaction medium, the rate of heat generated from the reaction, the heat absorption rate by the reaction mixture and the rate of heat removed from the reactor wall through conduction or convection.[92] Remarkably, the optical energy of photons that are not utilized in photochemical processes will be transformed to the thermal energy. An energy balance equation can be used to describe the heat transfer process without considering the effect of axial dispersion and the thermal energy produced by photons quenching:

$$\frac{dT}{dV} = \frac{dT}{S \cdot dz} = \frac{r \cdot (-\Delta H_R)}{mC_P} - \frac{U \cdot a \cdot (T - T_c)}{mC_P} \tag{21}$$

where T is the temperature of reaction mixture, V is the reactor volume, S is the cross-sectional area, z is the reactor length along the flow direction, a is the specific heat-exchange surface, ΔH_R is the reaction enthalpy, C_P is the mean specific heat capacity of the reaction mixture, and U is the global heat-transfer coefficient. The first item on the left-hand side describes the variation of the temperature along the flow direction. The first item on the right-hand side describes the rate of heat generated through reaction and the second item represents the rate of heat removed through the channel wall. Assuming that the cross-section is constant and the specific heat-exchange surface is equal to $4/d_h$ (d_h is the hydraulic diameter of microreactors), equation (4) can be rewritten as follows[93,94]:

$$\frac{dT}{dz} = \frac{1}{\rho u C_P} \left[r\Delta H_R - \frac{4U(T - T_c)}{d_h} \right] \tag{22}$$

In thermal processes, this equation cannot be solved analytically because the reaction rate (r) is controlled by the reaction temperature according to the Arrhenius equation. However, the reaction rate in photochemical processes is mainly controlled by the local volumetric rate of energy absorption and the quantum yield. A numerical solution of a typical concentration and temperature development

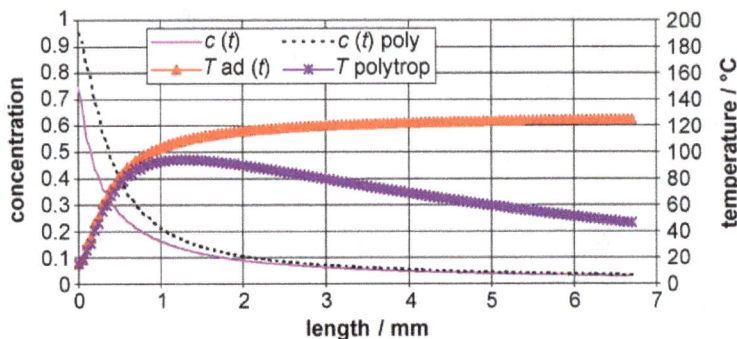

Fig. 2.7. Concentration and temperature development along the microchannel for a model exothermic reaction. (Reprinted with permission from Ref. 94. Copyright © 2008 John Wiley & Sons Inc.)

of an exothermic reaction along the reaction channel is shown in Figure 2.7.[94]

Apparently, the temperature profile in a reactor is highly dependent on the global heat transfer coefficient (U). It includes all heat transfer resistances and can be expressed as follows[95]:

$$\frac{1}{U} = \frac{1}{h_{in}} + \left(\frac{d_{out} - d_h}{2\lambda_{wall}}\right)\frac{a}{a_m} + \frac{1}{h_{out}}\frac{a}{a_{out}} \tag{23}$$

The first item in the right side describes the resistance between the channel wall and reaction mixture, the second item represents the resistance of channel wall, and the third item expresses the resistance located between the outer wall and the cooling fluid. The convective heat-transfer coefficients (h_{in} and h_{out}) are related to the hydrodynamics and the hydraulic diameter of reactors, which can be calculated through the Nusselt number (Nu)[96,97]:

$$h_{in} = Nu \cdot \lambda_{fluid}/d_h = \left[3.65 + \frac{0.19(Re \cdot Pr \cdot d_h)^{0.8}}{1 + 0.117(Re \cdot Pr \cdot d_h)^{0.467}}\right] \cdot \lambda_{fluid}/d_h \tag{24}$$

According to this correlation, a reduction in the hydraulic diameter (d_h) will result in a higher heat-transfer coefficient. This immediately rationalizes the higher heat transfer rate in microreactors compared with conventional reactors. Microheat exchangers with heat transfer efficiency even up to 790 W/cm^2 have been developed in the 1980s, which indeed enlightened the occurrence and development of microreactor technology.[98,99] In addition, higher conductivities of channel materials will result in a higher global heat transfer coefficient. However, the requirement of transparency significantly limits the types of materials which can be applied for the fabrication of photomicroreactors. The reactor materials should allow for efficient photon transport from the light source to the reaction medium. Various polymer materials, such as polymethylmethacrylate (PMMA), polydimethylsiloxane (PDMS), PFA, and FEP, are used to fabricate the photomicroreactors, due to their high light transmission, low cost and good compatibility with reaction environment.[100,101] Among these polymer materials, PFA and FEP materials are highly electrical insulating and flame retardant, self-lubricating and highly resistant to high temperature, chemical reagents and organic solvents. Capillary microreactors made of PFA and FEP materials through the special extrusion and sintering process are frequently applied in photochemical processes, which are wrapped around the lamps to maximize the photon flux. However, one also should consider the thermal conductivities of materials especially when performing photochemical processes with large reaction enthalpy. Table 2.3 summarizes the physical properties of materials commonly applied for fabrication photomicroreactors.[102,103] It is worth noting that the chemical resistance of a polymer material is mainly determined by the chemical structure of the material and the strength of the weakest link in this specific structure.

Special attentions should be paid to the extra heat produced by the light source. Typically, a cooling system has to be integrated to control the reaction temperature and avoid superheating of the light source that would lead to permanent damage of the entire photoreactor system. Inner jackets in which coolants are introduced for cooling are frequently applied in conventional photoreactors. For some compact photomicroreactor systems or the photomicroreactor systems with high-energy lamps, the cooling systems should be carefully designed to ensure effective cooling.[26] Kakiuchi *et al.* have

Table 2.3. Physical properties of materials commonly applied for fabrication of photomicroreactors.

Material	Thermal conductivity (λ) [W m^{-1} K^{-1}]	Transmission	Chemical compatibility
PFA	0.195	91–96% for visible light (400–700 nm) and 77–91% for UV light (250–400 nm) with 2–5 mm thickness	Highly flexible and resistant on strong acidic and alkaline solvents at moderate temperatures and pressures
FEP	0.19–0.24	FEP has similar properties for visible light and a bit better for UV light than PFA	Similar to FEP
PMMA	0.19	88% (>400 nm) with 2 mm thickness	PMMA and PDMS materials are easy to swell when subjected to many organic solvents
PDMS	0.15	>95% within visible light range with 3 mm thickness	Many solvents swell PDMS, such as dimethyl sulfoxide, acetonitrile diisopropylamine, triethylamine, pentane, and xylenes
Glass	1	About 95% for Pyrex (>275 nm), Corex (>260 nm), and Vycor (>220 nm) with 2 mm thickness	Resistant to most solvents at low to medium temperatures and it is crisp
Silicon	149	55% (>190 nm and <6 μm) with 5 mm thickness	Resistant on high temperatures and high pressures, a broad range of chemicals.
Silicon Carbide	120–490	Similar to silicon	Similar to silicon

reported on the gas–liquid [2 + 2] photocycloaddition of a chiral cyclohexenone with ethylene gas (Scheme 4) in photomicroreactors within the segmented flow regime.[104] The reactor consisted of FEP microcapillary (1.0 mm i.d.) coiled around a quartz immersion well in which a 500-W high-pressure Hg lamp was placed. The photomicroreactor was placed in a methanol cooling bath. Full conversion was obtained in 1 min with a selectivity of 52% *de*. The authors found that a rigorous cooling of the microreactor was necessary to obtain these high diastereoselectivities. The importance of efficient cooling with high-energy Hg lamps was also demonstrated by Jensen *et al.*[105] Without active cooling, temperatures in the microreactor could raise dramatically up to 250°C.[106,107]

Moreover, some intermediates/products in photochemical processes are highly thermally unstable and can undergo vigorous decomposition at elevated temperatures. The Wolff rearrangement involves the conversion of an α-diazocarbonyl compound into a ketene by loss of dinitrogen and accompanied with a 1,2-rearrangement.[108] Trans-β-lactams was produced in a capillary microreactor setup via an intramolecular Wolff rearrangement of α-diazo-β-ketoamides. Especially, external cooling of the capillary microreactor was required to avoid thermal decomposition of the diazocompounds. For the photocatalytic aerobic oxidation of thiols to disulfides with Eosin Y as a catalyst in a capillary photomicroreactor system, the temperature could be kept constant (room temperature) during the reaction process since high flow rates of pressurized air were used to cool the whole photomicroreactor system.[26,109] In addition, temperature probes such as thermocouples can be placed on/around the photomicroreactor walls, which can detect the temperature that is close to the real temperature inside reactors. The real-time temperature profile in the whole photomicroreactor system can be measured *in situ* by using an infrared camera.[110,111]

2.6 Conclusion

In this chapter, we have focused on giving an overview of transport phenomena associated with photochemical processes conducted in microreactors, which is beneficial for understanding the intrinsic advantages of microreactors in specific photochemical transformations. We have tried to explain relevant engineering principles in a language that is understandable for both synthetic chemists and chemical engineers. Some vivid examples have been given in order to understand the importance of transport properties inside reactors and certify the benefits of microreactors when they are applied for photochemical transformations. It is our hope that this chapter provides useful guiding ideologies in the engineering aspects especially related to transport properties when designing microreactors for photochemical processes.

In the past two decades, the characteristics of mixing, mass and heat transfer in microreactors with absence of any reactions or with the presence of thermal chemical reactions have been widely studied, and a number of correlations have been deduced to predict these transport properties.[1] As compared with the mass/heat transport, the photon transport phenomena are more complex. The energy transformation in photochemical processes involves electrical energy, optical energy and chemical energy, which is closely related to the transport phenomena. Moreover, most of correlations which have been derived for large-scale systems are only limitedly valid for the microreactors with specific structures. In particular, effects of entrance, surface roughness and wettability that are often neglected in conventional reactors can be highly relevant on a microscale. Therefore, more research should be carried out in order to comprehend the hydrodynamics and transport phenomena associated with photochemical processes in microreactors. Once the basic understanding of the transport phenomena and the intrinsic reaction characteristics is obtained, both mathematical modeling and numerical simulations can be applied to achieve proper photomicroreactor designs and process optimization.

The development of technologies such as measurement technique and information technology facilitates the study of mass transport phenomena in microreactors.[112,113] The direct measurement of the radiation distribution and spectral specific intensity inside microreactors depends on the miniaturization of light sources and photometers, which enables the exact and detailed description of photon transport phenomena in these microreactors. Special attentions should be taken in terms of the compatibility among light sources, microreactors and cooling systems. Furthermore, a holistic concept should be established to completely understand the relationship among hydrodynamics, transport properties and photochemical reaction performance in microreactors.

References

1. M. N. Kashid, A. Renken and L. Kiwi-Minsker, *Chem. Eng. Sci.* **66**, 3876 (2011).
2. V. Hessel, D. Kralisch, N. Kockmann, T. Noël and Q. Wang, *ChemSusChem* **6**, 746 (2013).
3. J. C. Pastre, D. L. Browne and S. V. Ley, *Chem. Soc. Rev.* **42**, 8849 (2013).
4. R. L. Hartman, J. P. McMullen and K. F. Jensen, *Angew. Chem. Int. Ed.* **50**, 7502 (2011).
5. T. Noël and S. L. Buchwald, *Chem. Soc. Rev.* **40**, 5010 (2011).
6. C. G. Frost and L. Mutton, *Green Chem.* **12**, 1687 (2010).
7. Y. H. Su, N. J. W. Straathof, V. Hessel and T. Noël, *Chem.-Eur. J.* **20**, 10562 (2014).
8. T. Bach and J. P. Hehn, *Angew. Chem. Int. Ed.* **50**, 1000 (2011).
9. N. Hoffmann, *Chem. Rev.* **108**, 1052 (2008).
10. O. Legrini, E. Oliveros and A. M. Braun, *Chem. Rev.* **93**, 671 (1993).
11. H. D. Roth, *Angew. Chem. Int. Ed.* **28**, 1193 (1989).
12. S. Chatani, C. J. Kloxin and C. N. Bowman, *Polym. Chem.* **5**, 2187 (2014).
13. G. P. Smestad and A. Steinfeld, *Ind. Eng. Chem. Res.* **51**, 11828 (2012).
14. O. M. Alfano, D. Bahnemann, A. E. Cassano, R. Dillert and R. Goslich, *Catal. Today* **58**, 199 (2000).
15. J. C. Andre, A. Tournier and X. Deglise, *J. Photochem.* **22**, 7 (1983).
16. D. Dworkin and J. S. Dranoff, *AIChE J.* **24**, 1134 (1978).
17. M. Fischer, *Angew. Chem. Int. Ed.* **17**, 16 (1978).
18. A. K. Suresh, M. M. Sharma and T. Sridhar, *Ind. Eng. Chem. Res.* **39**, 3958 (2000).
19. T. Noël, X. Wang and V. Hessel, *Chim. Oggi* **31**, 10 (2013).
20. M. Oelgemoller, *Chem. Eng. Technol.* **35**, 1144 (2012).
21. J. P. Knowles, L. D. Elliott and K. I. Booker-Milburn, *Beilstein J. Org. Chem.* **8**, 2025 (2012).
22. M. Oelgemoller and O. Shvydkiv, *Molecules* **16**, 7522 (2011).
23. Z. J. Garlets, J. D. Nguyen and C. R. J. Stephenson, *Isr. J. Chem.* **54**, 351 (2014).
24. E. E. Coyle and M. Oelgemoller, *Photochem. Photobiol. Sci.* **7**, 1313 (2008).

25. A. E. Cassano, C. A. Martin, R. J. Brandi and O. M. Alfano, *Ind. Eng. Chem. Res.* **34**, 2155 (1995).
26. Y. Su, A. Talla, V. Hessel and T. Noël, *Chem. Eng. Technol.* **38**, 1733 (2015).
27. T. Aillet, K. Loubière, O. Dechy-Cabaret and L. Prat, *Chem. Eng. Technol.* **39**, 115 (2016).
28. R. Gorges, S. Meyer and G. Kreisel, *J. Photochem. Photobiol. A.* **167**, 95 (2004).
29. S. Meyer, D. Tietze, S. Rau, B. Schaefer and G. Kreisel, *J. Photochem. Photobiol. A.* **186**, 248 (2007).
30. A. Jamali, R. Vanraes, P. Hanselaer and T. Van Gerven, *Chem. Eng. Process.* **71**, 43 (2013).
31. B. Shen, M. W. Bedore, A. Sniady and T. F. Jamison, *Chem. Commun.* **48**, 7444 (2012).
32. B. D. A. Hook, W. Dohle, P. R. Hirst, M. Pickworth, M. B. Berry and K. I. Booker-Milburn, *J. Org. Chem.* **70**, 7558 (2005).
33. K. G. Maskill, J. P. Knowles, L. D. Elliott, R. W. Alder and K. I. Booker-Milburn, *Angew. Chem. Int. Ed.* **52**, 1499 (2013).
34. T. Fukuyama, Y. Kajihara, Y. Hino and I. Ryu, *J. Flow. Chem.* **1**, 40 (2011).
35. Y. H. Su, V. Hessel and T. Noël, *AIChE J.* **61**, 2215 (2015).
36. T. Aillet, K. Loubiere, O. Dechy-Cabaret and L. Prat, *Chem. Eng. Process.* **64**, 38 (2013).
37. G. D. Harris, V. D. Adams, W. M. Moore and D. L. Sorensen, *J. Environ. Eng. ASCE* **113**, 612 (1987).
38. H. J. Kuhn, S. E. Braslavsky and R. Schmidt, *Pure. Appl. Chem.* **76**, 2105 (2004).
39. G. Gauglitz and S. Hubig, *J. Photochem.* **17**, 13 (1981).
40. S. Jin, A. A. Mofidi and K. G. Linden, *J. Environ. Eng. ASCE* **132**, 831 (2006).
41. T. Aillet, K. Loubiere, O. Dechy-Cabaret and L. Prat, *Int. J. Chem. React. Eng.* **12**, 257 (2014).
42. M. A. Cismesia and T. P. Yoon, *Chem. Sci.* **6**, 5426 (2015).
43. L. S. Mendez-Portillo, M. Heniche, C. Dubois and P. A. Tanguy, *AIChE J.* **59**, 988 (2013).
44. C. Iliescu, *Inf. MIDEM* **36**, 204 (2006).
45. S. Cerbelli and M. Giona, *Chem. Eng. J.* **138**, 523 (2008).
46. V. Hessel, S. Hardt, H. Lowe and F. Schonfeld, *AIChE J.* **49**, 566 (2003).
47. I. Akkerman, M. Janssen, J. Rocha and R. H. Wijffels, *Int. J. Hydrogen. Energy* **27**, 1195 (2002).
48. P. R. Harris and J. S. Dranoff, *AIChE J.* **11**, 497 (1965).
49. J. A. Davies, D. L. Boucher and J. G. Edwards, The question of artificial photosynthesis of ammonia on heterogeneous catalysts, in *Advances in Photochemistry*, Vol. **19**, Eds. D. C. Neckers, D. H. Volman and G. von Bunau (Wiley, 2007).

50. T. Van Gerven, G. Mul, J. Moulijn and A. Stankiewicz, *Chem. Eng. Process.* **46**, 781 (2007).
51. V. Hessel, H. Lowe and F. Schonfeld, *Chem. Eng. Sci.* **60**, 2479 (2005).
52. N. Kockmann, T. Kiefer, M. Engler and P. Woias, *Sens. Actuators B. Chem.* **117**, 495 (2006).
53. J. R. Bourne, *Org. Process. Res. Dev.* **7**, 471 (2003).
54. R. S. Schechter and E. H. Wissler, *Appl. Sci. Res.* **9**, 334 (1960).
55. O. M. Alfano, R. L. Romero and A. E. Cassano, *Chem. Eng. Sci.* **41**, 421 (1986).
56. O. M. Alfano, R. L. Romero and A. E. Cassano, *Chem. Eng. Sci.* **41**, 1137 (1986).
57. H. A. Irazoqui, J. Cerdá and A. E. Cassano, *Chem. Eng. J.* **11**, 27 (1976).
58. T. Aillet, K. Loubiere, L. Prat and O. Dechy-Cabaret, *AIChE J.* **61**, 1284 (2015).
59. N. J. Straathof, Y. Su, V. Hessel and T. Noël, *Nat. Protocol.* **11**, 10 (2016).
60. O. Shvydkiv, A. Yavorskyy, S. B. Tan, K. Nolan, N. Hoffmann, A. Youssef and M. Oelgemoller, *Photochem. Photobiol. Sci.* **10**, 1399 (2011).
61. A. Talla, B. Driessen, N. J. W. Straathof, L. G. Milroy, L. Brunsveld, V. Hessel and T. Noël, *Adv. Synth. Catal.* **357**, 2180 (2015).
62. P. O. Mchedlov-Petrossyan, G. A. Khomenko and W. B. Zimmerman, *Chem. Eng. Sci.* **58**, 3005 (2003).
63. J. G. van de Vusse, *Chem. Eng. Sci.* **21**, 631 (1966).
64. J. G. van de Vusse, *Chem. Eng. Sci.* **21**, 645 (1966).
65. J. G. van de Vusse, *Chem. Eng. Sci.* **21**, 1239 (1966).
66. P. R. Ogilby, *Chem. Soc. Rev.* **39**, 3181 (2010).
67. H. Mimoun, *Angew. Chem. Int. Ed.* **21**, 734 (1982).
68. D. R. Kearns, *Chem. Rev.* **71**, 395 (1971).
69. F. Levesque and P. H. Seeberger, *Org. Lett.* **13**, 5008 (2011).
70. Y. H. Su, G. W. Chen, Y. C. Zhao and Q. Yuan, *AIChE J.* **55**, 1948 (2009).
71. Y. H. Su, Y. C. Zhao, G. W. Chen and Q. A. Yuan, *Chem. Eng. Sci.* **65**, 3947 (2010).
72. Y. C. Zhao, G. W. Chen and Q. Yuan, *AIChE J.* **53**, 3042 (2007).
73. J. Tan, Y. C. Lu, J. H. Xu and G. S. Luo, *Chem. Eng. J.* **181**, 229 (2012).
74. D. Tsaoulidis and P. Angeli, *Chem. Eng. J.* **262**, 785 (2015).
75. J. Yue, G. W. Chen, Q. Yuan, L. G. Luo and Y. Gonthier, *Chem. Eng. Sci.* **62**, 2096 (2007).
76. F. K. Kies, B. Benadda and M. Otterbein, *Chem. Eng. Process.* **43**, 1389 (2004).
77. E. Dluska, S. Wronski and T. Ryszczuk, *Exp. Thermal. Fluid Sci.* **28**, 467 (2004).
78. A. Heyouni, M. Roustan and Z. Do-Quang, *Chem. Eng. Sci.* **57**, 3325 (2002).
79. D. Herskowits, V. Herskowits, K. Stephan and A. Tamir, *Chem. Eng. Sci.* **45**, 1281 (1990).
80. J. C. Charpentier, *Adv. Chem. Eng.* **11**, 1 (1981).
81. C. X. Zhao and A. P. J. Middelberg, *Chem. Eng. Sci.* **66**, 1394 (2011).

82. C. Schweitzer and R. Schmidt, *Chem. Rev.* **103**, 1685 (2003).
83. K. I. Salokhiddinov, I. M. Byteva and G. P. Gurinovich, *J. Appl. Spectrosc.* **34**, 561 (1981).
84. L. A. Belfiore, J. J. Way and Z. Li, *Transport Phenomena for Chemical Reactor Design* (Wiley, 2006).
85. L. Kucka, J. Richter, E. Y. Kenig and A. Gorak, *Sep. Purif. Technol.* **31**, 163 (2003).
86. J. Fischer, T. Lange, R. Boehling, A. Rehfinger and E. Klemm, *Chem. Eng. Sci.* **65**, 4866 (2010).
87. C. Kutal, *J. Chem. Educ.* **60**, 882 (1983).
88. S. Linic, U. Aslam, C. Boerigter and M. Morabito, *Nat. Mater.* **14**, 744 (2015).
89. Govindjee, *Crop. Sci.* **7**, 551 (1967).
90. D. S. Meyaard, Q. F. Shan, J. Cho, E. F. Schubert, S. H. Han, M. H. Kim, C. Sone, S. J. Oh and J. K. Kim, *Appl. Phys. Lett.* **100**, 081106 (2012).
91. Y. S. Mimieux Vaske, M. E. Mahoney, J. P. Konopelski, D. L. Rogow and W. J. McDonald, *J. Am. Chem. Soc.* **132**, 11379 (2010).
92. M. N. Kashid, A. Renken and L. Kiwi-Minsker, *Microstructured Devices for Chemical Processing* (Wiley-VCH, Weinhein, 2014).
93. D. M. Roberge, B. Zimmermann, F. Rainone, M. Gottsponer, M. Eyholzer and N. Kockmann, *Org. Process. Res. Dev.* **12**, 905 (2008).
94. N. Kockmann, M. Gottsponer, B. Zimmermann and D. M. Roberge, *Chem.-Eur. J.* **14**, 7470 (2008).
95. J. Haber, M. N. Kashid, A. Renken and L. Kiwi-Minsker, *Ind. Eng. Chem. Res.* **51**, 1474 (2012).
96. T. L. Bergman and F. P. Incropera, *Fundamentals of Heat and Mass Transfer*, 7th Edn. (Wiley, Hoboken, NJ, 2011).
97. W. M. Deen, *Analysis of Transport Phenomena* (Oxford University Press, New York, 1998).
98. Y. M. Joshi and S. Khandekar, Nanoscale and microscale phenomena: Fundamentals and applications, in *Tracts in Mechanical Engineering* (Springer, 2015).
99. J. J. Brandner, L. Bohn, T. Henning, U. Schygulla and K. Schubert, *Proc. 4th Int. Conf. Nanochannels, Microchannnels, and Minichannels, Points A and B*, 2006, pp. 1233–1243.
100. J. P. McMullen and K. F. Jensen, *Ann. Rev. Anal. Chem.* **3**, 19 (2010).
101. M. J. Madou, *Fundamentals of Microfabrication: The Science of Miniaturization*, 2nd edn. (CRC Press, Boca Raton, FL, 2002).
102. J. N. Lee, C. Park and G. M. Whitesides, *Anal. Chem.* **75**, 6544 (2003).
103. Y. C. Chen, W. L. Tseng and C. H. Lin, *Int. J. Autom. Smart Technol.* **2**, 63 (2012).
104. K. Terao, Y. Nishiyama, H. Tanimoto, T. Morimoto, M. Oelgemoller and K. Kakiuchi, *J. Flow. Chem.* **2**, 73 (2012).
105. K. Pimparkar, B. Yen, J. R. Goodell, V. I. Martin, W. H. Lee, J. A. Porco, A. B. Beeler and K. F. Jensen, *J. Flow. Chem.* **1**, 53 (2011).

106. M. Nettekoven, B. Pullmann, R. E. Martin and D. Wechsler, *Tetrahedron Lett.* **53**, 1363 (2012).
107. A. Vasudevan, C. Villamil, J. Trumbull, J. Olson, D. Sutherland, J. Pan and S. Djuric, *Tetrahedron Lett.* **51**, 4007 (2010).
108. W. Kirmse, *Eur. J. Org. Chem.* **2002**, 2193 (2002).
109. A. Talla, B. Driessen, N. J. W. Straathof, L. G. Milroy, L. Brunsveld, V. Hessel and T. Noël, *Adv. Synth. Catal.* **357**, 2180 (2015).
110. T. Illg, V. Hessel, P. Lob and J. C. Schouten, *Green Chem.* **14**, 1420 (2012).
111. T. Illg, V. Hessel, P. Lob and J. C. Schouten, *ChemSusChem* **4**, 392 (2011).
112. C. Q. Yao, Z. Y. Dong, Y. C. Zhao and G. W. Chen, *Chem. Eng. Sci.* **112**, 15 (2014).
113. S. Fransen and S. Kuhn, (2016). *React. Chem. Eng.* **1**, 288.

Chapter 3

Modeling of Photochemical Processes in Continuous-Flow Reactors

Anca Roibu and Simon Kuhn*

*Department of Chemical Engineering, KU Leuven
Celestijnenlaan 200F, 3001 Leuven, Belgium
simon.kuhn@kuleuven.be

3.1 Introduction

Despite the progress in the past decades in the area of lightdriven reactions, only a limited number of photochemical reactions is implemented at large scale.[1,2] The major hurdle which needs to be overcome is linked to the scale-up strategy.[3] The limited light penetration (described by Lambert–Beer's law) rapidly decreases the transformation efficiency in large-scale batch reactors as only non-uniform light intensities are achieved in the vessel. However, using continuous-flow reactors provides an avenue for efficient scale-up. Moreover, the characteristic reactor sizes on the micro or milliscale allow for a homogeneous light distribution in the reacting volume and advanced designs ensure sufficient productivity at the desired scale.[4]

Another reason is the difficulty of modeling photochemical reactors.[3] Scaling up can be realized by starting working with a laboratory-scale reactor and continue by gradually increasing the reactor size until reaching the desired productivity or by predicting the performance of the large-scale reactor with mathematical modeling. Modeling is more precise and cost efficient, but is characterized by a higher complexity.[5,6] Consequently, most pilot-scale photoreactors are still designed using empirical or semi-empirical methods.[7] Predicting the performance of a large-scale reactor of a different geometry than the one investigated in the laboratory requires knowing the intrinsic kinetic parameters of the reaction.[6,8] Alfano and Cassano described a methodology for scale-up photoreactors which uses modeling at small and large scale: firstly for determination of the kinetic parameters and then for predicting the pilot-scale reactor performance.[8]

This chapter aims to introduce the main concepts of a holistic scale-up strategy based on combined modeling and experiments. We will highlight the important steps of photoreactor modeling such as the kinetic model, radiation, mass and momentum balance, etc., and which are the variables to be exchanged between them to ensure efficient and robust scaling. Furthermore, examples from various application fields of photochemistry, e.g. water treatment and synthetic chemistry, will be offered to illustrate the implementation of the described concepts. This chapter will appeal to both chemical engineers and chemists aiming to develop modeling as a tool for scaling-up photochemical reactors.

3.2 Modeling Continuous Photochemical Processes for Photoreactor Scaling Up

Alfano and Cassano reported a scale-up methodology based on photoreactor modeling by coupling the radiation, mass and momentum balances with the kinetic model.[8] Figure 3.1 presents the main steps of this method and was adapted from the flow scheme reported by Ghafoori.[6]

The scale-up approach encompasses two parts: one involves using a laboratory-scale reactor and the second a large-scale reactor. The laboratory-scale reactor is used to determine the intrinsic kinetic parameters; their use allows the possibility to implement the reaction in a photoreactor of a different size, geometry and operated under different conditions.[6] The first step is to determine the mechanism of the investigated reaction, as the kinetic model is designed based

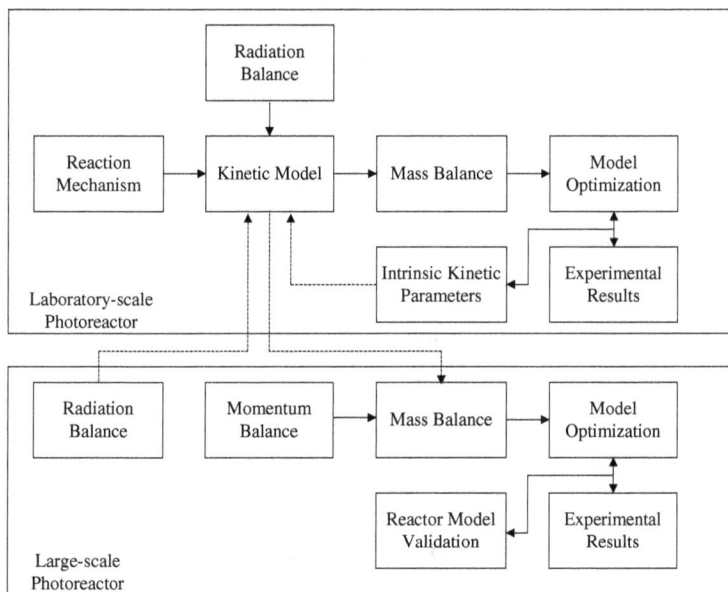

Fig. 3.1. Scale-up methodology of photoreactors. (Adapted from Ref. 6. Copyright © 2014 Elsevier B.V.)

on the reaction scheme. Alfano and Cassano recommend to have a validated kinetic model and highlight the importance of fundamental chemistry research to obtain reliable results.[8] In the next steps, the amount of absorbed photons determined from the radiation balance is included in the kinetic model, which is integrated further in the material balance. The intrinsic parameters are estimated after optimizing the algorithm, so the errors between the predictions of the model and the experimental results are minimized.

Once the kinetic parameters are determined, they are included in the same kinetic model which is coupled with the radiation balance characteristic of the large-scale reactor. The reaction rate determined in this way is integrated in the mass balance together with the momentum balance. The momentum balance is not present in the case of the laboratory-scale reactor because it is recommended to employ a well-mixed reactor which guarantees uniform reactant concentration in the entire volume or in the cross-sections of the reactor.

The balances realized for the large-scale reactor are solved simultaneously, optimized and validated with the experimental results.[6]

The following aspects need to be considered when applying this method[8]:

- a validated reaction scheme is required
- operation of the laboratory-scale reactor with no mass-transfer limitations
- the same catalyst properties (preparation protocol and morphology) in the case of heterogeneous catalytic system
- isothermal conditions
- light sources with the same spectral radiation output
- application and rigorously solving the mathematical model for the laboratory- and large-scale reactors

Applications of this scale-up methodology are briefly described in Section 3.7.

3.3 Kinetic Model

The existence of the radiation-activated step distinguishes a photochemical from a thermal reaction.[9] In a single photon absorption photochemical reaction (which we define here to include both photosensitized and photocatalytic reactions), the rate of this initiation step (primary event), r, is proportional to the local volumetric rate of photon absorption (LVRPA)[10,11]:

$$r = \Phi_\lambda e_\lambda^a \tag{1}$$

where Φ is the reaction quantum yield and e^a is the LVRPA. Both properties are defined at a given wavelength, λ, and for a given species.[11]

The LVRPA is dependent on the radiation field in the reactor and can be determined by solving the radiation balance which will be described in Section 3.4.[12]

The reaction quantum yield in this case is called primary quantum yield.[10] A primary reaction includes the absorption of the photon which leads to the formation of an excited state followed by the formation of an intermediate or final product and/or deactivation processes (e.g. non-radiative, radiative: fluorescence, phosphorescence).[13] However, it excludes the thermal reactions of the intermediate or product which may take place in the reactor space. As the primary quantum yield considers a single photon absorption process, its value will be comprised between 0 and 1. A value of 0 will indicate that only deactivation processes occur and a value of 1 that no deactivation processes are present (quantitative conversion of the excited state molecules to the product). In some cases, the quantum yield is reported to be higher than 1 (e.g. photochemical polymerization reactions). However, this measure represents the overall quantum yield and considers both photochemical and thermal reactions.[13] Care must be taken when using this value, as it does not have the quality of an intrinsic kinetic property.[10]

In heterogeneous catalytic systems it is difficult to rigorously determine the number of absorbed photons, because a large amount of photons is lost through scattering. Therefore, the quantum yield can be determined only after quantifying the scattering effects (a method to determine the optical properties of suspended catalyst is described by Alfano *et al.* in Chapter 4). However, one needs to pay attention because most of the investigations report the apparent quantum yield and not the "true" quantum yield. The former property is defined as the ratio of the rate of the reactant consumed or product formed and amount of incident photons.[14]

Establishing the kinetic model is a demanding process especially in the case of reactions with complex mechanism. As a validated kinetic model is required, the scale-up investigations are generally based on extensive kinetic studies of the reactions to be implemented at large-scale.[6,15−18]

3.4 Radiation Balance

3.4.1 *Radiative transfer equation*

The requirement of LVRPA determination leads to the coupling of the photochemical reaction rate with the radiation balance. This property can be calculated from the radiation intensity integrated over all propagation directions as follows[12,19]:

$$e^a = \int_{\lambda_1}^{\lambda_2} \kappa_\lambda(\boldsymbol{x}) \int_{\Omega=4\pi} I_{\lambda,\boldsymbol{\Omega}}(\boldsymbol{x}) d\Omega d\lambda \qquad (2)$$

where κ_λ is the volumetric absorption coefficient, $I_{\lambda,\boldsymbol{\Omega}}$ is the spectral intensity (reported also as specific spectral intensity), λ is the radiation wavelength, $\boldsymbol{x}$ is a position vector in a three-dimensional space, Ω is the solid angle of radiation propagation and $\boldsymbol{\Omega}$ is the unit vector in the direction of radiation propagation.

The spectral intensity can be obtained by solving the equation which describes the transport of radiation through the reacting medium (radiative transport equation [RTE]). By assuming negligible radiation emission due to the low operating temperature, steady state, elastic scattering (i.e. scattering changes only the direction of the photons), multiple and independent scattering, the RTE can be expressed as[12,19]:

$$\frac{dI_{\lambda,\boldsymbol{\Omega}}(\boldsymbol{x})}{ds} = -\kappa_\lambda(\boldsymbol{x})I_{\lambda,\boldsymbol{\Omega}}(\boldsymbol{x}) - \sigma_\lambda(\boldsymbol{x})I_{\lambda,\boldsymbol{\Omega}}(\boldsymbol{x})$$

$$+ \frac{\sigma_\lambda(\boldsymbol{x})}{4\pi} \int_{\Omega'=4\pi} p(\boldsymbol{\Omega'} \to \boldsymbol{\Omega}) I_{\lambda,\boldsymbol{\Omega'}}(\boldsymbol{x}) d\Omega' \qquad (3)$$

where σ_λ is the volumetric scattering coefficient, $p(\boldsymbol{\Omega'} \to \boldsymbol{\Omega})$ is the scattering phase function, $\boldsymbol{\Omega'}$ is the new solid angle of radiation propagation after scattering and s is the linear coordinate along the direction of radiation propagation $\boldsymbol{\Omega}$.

The RTE is an integro-differential equation, where the first term on the right-hand side quantifies the loss of radiation through absorption, the second term, the loss of radiation due to out-scattering and the third term the radiation gained as a consequence of in-scattering (resulting from multiple scattering phenomena).[12] The equation has to be solved for each wavelength, λ, of the discretized wavelength range employed during the photochemical reaction.[19] A detailed description of the parameters involved in the RTE is offered by Alfano *et al.* in Chapter 4.

3.4.2 *Numerical resolution of radiative transport equation*

The RTE is simplified to the Bouguer–Lambert–Beer law when scattering is negligible i.e. the radiation travels through a homogeneous medium. This case is described in more detail by Su *et al.* in Chapter 2. However, it is difficult to analytically solve the equation when it is applied to heterogeneous systems. Solving the RTE involves appropriate boundary conditions and numerical methods. The boundary conditions for the RTE depend on the type of emission of the lamp and the geometrical properties of the lamp-reactor system.[10] Lamp emission models were reviewed previously,[9,20] and will not be discussed in this chapter. The RTE solution can be calculated using discretization, such as the discrete-ordinate method (DOM), by probabilistic numerical methods such as Monte Carlo (MC) or by simplified radiation field models, such as two-flux method (TFM) and six-flux method (SFM).[19,21] The main methods used for solving the RTE will be briefly introduced from the most rigorous to the most simplified methods.

The DOM employs a directional and spatial discretization which transforms the RTE into a system of algebraic equations.[22] These equations describe the transport of photons in such a way that they can be solved following the direction of propagation, starting from the values provided by the boundary conditions.[19] The DOM is recognized as one of the most accurate methods, because it does not involve any assumption, other than the discretization of pertinent partial different equations.[23,24] However, the application of the DOM for its solution requires a high degree of numerical accuracy, which is reflected in a high computational time.[25] As the DOM does not guarantee the conservation of the radiant energy at the surfaces of complex geometries especially for anisotropic scattering, a variant DO model based on the finite-volume method (FVM) was developed. FVM allows to conserve radiative energy, as the RTE is integrated over both the control angle and the control volume, unlike the DOM, in which the RTE is integrated over the control volume only.[24]

The MC method is based on the probabilistic behavior of photons, since they are emitted from the light source until their absorption or escape through the reactor wall.[22] Firstly, the position and the direction of the photons are randomly determined taking into account the geometry (e.g. the position of the light source). Then, the photons travel through the medium and can be absorbed or scattered in a certain direction. The probabilities of being absorbed or scattered depend on the absorption and scattering coefficients. For the scattered photon, a new direction is determined whose probability depends on the phase function.[13] The MC method has the advantage that it is relatively simple to apply, especially for complex geometries.[26] As it can be observed in Table 3.1, MC simulations were successfully applied to annular reactors, monolith reactors, multiplate reactors, etc. Moreover, this method is considered to be precise, as no approximations are included regarding the radiation field; its precision is determined by the number of generated photons and their stochastic behavior. As a consequence, the MC method can easily become computationally expensive.[20,27]

The TFM does not use the RTE in an explicit manner. The incident radiation is divided into two opposite directions and implies two coupled ordinary differential equations for which analytical solutions can be easily determined.[20] It is based on the assumption that the scattered photons are scattered only in backward direction. It is probably the simplest method to study the light intensity distribution in heterogeneous systems.[20] The SFM is a three-dimensional extension of the TFM, so the photons are scattered in six principal directions with respect to the incoming radiation.[28] Its mathematical structure is of algebraic nature. Therefore, its implementation in reactors of different scales and radiation sources implies low complexity numeric procedures and short computational times.[25] Moreover, the SFM has the advantage of offering a quite accurate representation of the radiation field while using simple modeling equations.[29,30]

In the following we highlight the difference in results obtained for different levels of model complexity. Li Puma and Brucato compared the relative error between experimental results and three models,

Table 3.1. Applications of the main methods used for RTE resolution.

Method	Reactor configuration	References
DOM	Annular	23
		31
	Cylindrical, rectangular	5
	Cylindrical	19
FVM	Annular	24
	Single and multiple-lamp reactors	32
MC	Annular	33
		34
		35
		36
		37
	Plane-slab	29
	Monolyth	38
	Multiplate	26
	Cylindrical	39
TFM	Plane-slab	29
	Annular	21
	Thorus-shaped photobioreactor	40
	Flat-plate	41
	Microreactor	13
	Tubular	25
SFM	Annular	28
		21
		7
		30
	Plane-slab	29
	Falling film	7
	Compound parabolic collectors (CPC)	42
	Tubular	25

namely Lambert–Beer, TFM and SFM.[21] The experimental results were acquired for isoproturon oxidation in aqueous suspensions of TiO_2 in an annular photocatalytic reactor operated in recirculation mode. The Lambert–Beer model neglects scattering within the reactor space, therefore it overestimates the amount of radiation available for the photoreaction. As it can be observed in Figure 3.2,

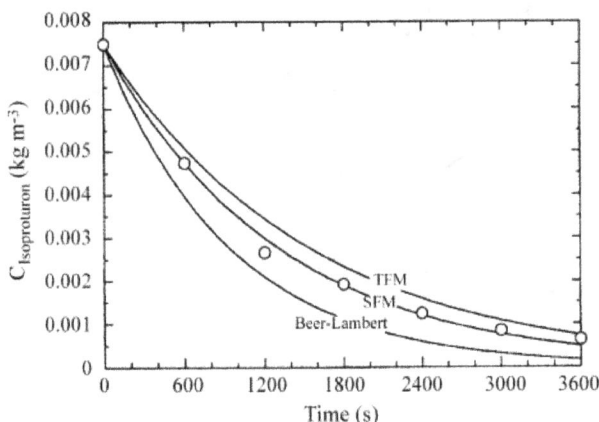

Fig. 3.2. Modeling the photocatalytic oxidation of isoproturon using Beer–Lambert model, TFM and SFM. Lines are predicted results and circles the experimental results. (Reprinted with permission from Ref. 21. Copyright © 2007 Elsevier B.V.)

the isoproturon conversion simulated using the Lambert–Beer model is higher than the conversion achieved in the experiments. On the other hand, the TFM considers back-scattering only (no contribution from in-scattering phenomena). Consequently, the results predicted by TFM underestimate the conversion of the herbicide. The study of Li Puma and Brucato clearly shows that (a) the experimental results will be between the results predicted by the two models, Beer–Lambert and TFM; (b) the radiation scattering should be considered in order to avoid the erroneous determination of the kinetic parameters, which are essential for the scale-up studies.[21]

For further examples of applications of the different modeling strategies to various reactor geometries the reader is referred to the selected references in Table 3.1.

3.5 Mass Balance

The mass balance for a component i in a reactor considering non-steady state is[43]:

$$\frac{\partial C_i}{\partial t} = -(\nabla \cdot C_i \mathbf{v}) - (\nabla \cdot J_i) + R_i \qquad (4)$$

where C_i is the molar concentration of species i, $\mathbf{v}$ is the velocity vector of the fluid, J_i is the molar diffusion flux and R_i is the volumetric rate of generation of component i. The kinetic model which includes the LVRPA is incorporated in the mass balance through the reaction rate R_i.

$\partial C_i / \partial t$ represents the rate of increase of component i per unit volume, $(\nabla \cdot C_i \mathbf{v})$ is the net rate of addition of i per unit volume by convection, $(\nabla \cdot J_i)$ is the net rate of addition of i per volume by diffusion and R_i is the net rate of production of i per unit volume by reaction.[43] According to Fick's law, the diffusive flux for a constant diffusivity of component i, D_i, is[43]:

$$J_i = -D_i(\nabla C_i) \tag{5}$$

and the mass balance for component i becomes[43]:

$$\frac{\partial C_i}{\partial t} = -(\nabla \cdot C_i \mathbf{v}) + D_i(\nabla^2 \cdot C_i) + R_i \tag{6}$$

The mass balance can be expressed in rectangular coordinates (to be used in the case of flat plate reactors, for example):

$$\frac{\partial C_i}{\partial t} = -\left(v_x \frac{\partial C_i}{\partial x} + v_y \frac{\partial C_i}{\partial y} + v_z \frac{\partial C_i}{\partial z} \right)$$
$$+ D_i \left(\frac{\partial^2 C_i}{\partial x^2} + \frac{\partial^2 C_i}{\partial y^2} + \frac{\partial^2 C_i}{\partial z^2} \right) + R_i \tag{7}$$

or in cylindrical coordinates (for annular and cylindrical reactors):

$$\frac{\partial C_i}{\partial t} = -\left(v_r \frac{\partial C_i}{\partial r} + v_\theta \frac{1}{r} \frac{\partial C_i}{\partial \theta} + v_z \frac{\partial C_i}{\partial z} \right)$$
$$+ D_i \left(\frac{1}{r} \frac{\partial}{\partial r} \left(r \frac{\partial C_i}{\partial r} \right) + \frac{1}{r^2} \frac{\partial^2 C_i}{\partial \theta^2} + \frac{\partial^2 C_i}{\partial z^2} \right) + R_i \tag{8}$$

In the following subsections, we will show possible simplifications of equations (7) and (8) in various cases. The material balance will be discussed for laboratory-scale reactors where it was used for determining the intrinsic kinetic parameters and for large scale reactors where it provided the tool to predict the reactor performance.

3.5.1 *Mass balance in laboratory scale reactor*

The general approach in kinetic investigations is to bring the studied system in the reaction rate limited regime, so the apparent reaction rate can reach the intrinsic value.[44] The mass balance is used to predict the conversion of a species, i, which is then compared with the experimental measurements. The unknown intrinsic kinetic parameters are obtained with a nonlinear parameter estimator and an optimization program. The final outcome is a kinetic model with known constants that can be used for scale-up purposes and reactor optimization, as this result is independent of the employed light source, particular reactor geometry and operating conditions.[12]

3.5.1.1 *Mass balance in batch reactor with recirculation*

The classical experimental setup used for kinetic studies is composed of a laboratory-scale reactor and a mixing tank. In the following part we describe possible simplifications of the general mass balance and in which conditions they can be considered.

Labas *et al.* studied the kinetics of oxidation of dilute aqueous formic acid solutions with UV radiation and H_2O_2.[45] The kinetic parameters were obtained in a well-stirred, small, batch flat-plate photoreactor. The authors considered the following conditions:

- the entire system operates under well-stirred conditions (well mixed)
- the ratio of the reactor volume ($70\,cm^3$) to the tank volume ($2000\,cm^3$) is <1 (recommended to be ≪1)
- the recirculating flow rate is high in order to have differential conversion per pass in the photoreactor

Since the reactor content was well mixed, the net rate of addition of i per unit volume by convection and by diffusion was assumed negligible and the mass balance could be written as

$$\frac{\partial C_i(t)}{\partial t} = R_{i,\lambda} \tag{9}$$

Equation (9) was then integrated over the system volume. The integral of the left-hand side could be divided in two volumes: the photoreactor volume (V_R) and the remaining system volume $(V_T - V_R$, where V_T is the total volume). The reaction is considered to only take place in the reactor, which yields:

$$\int_{V_R} \frac{\partial C_i(t)}{\partial t} dV + \int_{V_T - V_R} \frac{\partial C_i(t)}{\partial t} dV = \int_{V_R} R_{i,\lambda} dV \qquad (10)$$

As the volumes are constant, the derivative and integral signs can be interchanged. To each resulting term, the volume-average theorem is applied with $\langle\rangle$ indicating the volume-averaged value.

$$V_R \frac{\partial}{\partial t} \langle C_i(t)\rangle_{V_R} + (V_T - V_R) \frac{\partial}{\partial t} \langle C_i(t)\rangle_{V_T - V_R} = V_R \langle R_{i,\lambda}(x,t)\rangle_{V_R} \qquad (11)$$

Equation (11) can be divided by V_T and rearranged as follows:

$$\frac{V_R}{V_T} \frac{\partial}{\partial t} [\langle C_i(t)\rangle_{V_R} - \langle C_i(t)\rangle_{V_T - V_R}] + \frac{\partial}{\partial t} \langle C_i(t)\rangle_{V_T - V_R}$$

$$= \frac{V_R}{V_T} \langle R_{i,\lambda}(x,t)\rangle_{V_R} \qquad (12)$$

Because V_R/V_T was small ($V_R = 70\,\text{cm}^3$ and $V_T = 2000\,\text{cm}^3$) and the conversion per pass is differential (difference between outlet and inlet concentrations in the reactor was small) the first term of equation (12) was considered negligible. Consequently, when the system was well-mixed, the change in concentrations in $V_T - V_R$ was equal to the changes in concentrations in the tank as written in equation (13):

$$\left.\frac{dC_i(t)}{dt}\right|_{\text{Tank}} = \frac{V_R}{V_T} \langle R_{i,\lambda}(x,t)\rangle_{V_R} \qquad (13)$$

The reaction rate averaged over the reactor volume was estimated using the kinetic model and then compared with the reaction rate determined experimentally (evolution of formic acid concentration with time). The reaction rate was a function of position (x) because of the light distribution inside the reaction space. The kinetic parameters obtained in the described system were applied in the modeling of a large-scale reactor. The predicted concentrations were in reasonably good agreement with the measured concentrations. The errors found when running the large-scale reactor under different conditions were comprised between 6.6% and 11.8%.[45]

In kinetic studies, the predominantly used photoreactor geometries which are operated in recirculation regime are the annular reactor,[6,7,30,46] flat-plate reactor[16,45,47–50] and cylindrical reactor.[5,19,51]

3.5.1.2 *Mass balance in microreactors*

Recently, microreactors have been used for the determination of the kinetic constants of photocatalytic reactions.[44,52–54] Microreactors offer the advantage of a higher spatial illumination homogeneity and better light penetration through the reactor depth in comparison to other laboratory-scale reactors which are generally employed for kinetic studies.

When considering the mass balance for microreactors, usually a plug-flow reactor model in the absence or presence of mass transport limitations is assumed. The mass balance for a plug-flow reactor in the presence of radial concentration profiles and axial dispersion in two-dimensional rectangular coordinates (x, y) can be expressed as (the channel length corresponds to the x-direction and the channel width to the y-direction):

$$v_x \frac{dC_i}{dx} - D_i \frac{d^2 C_i}{dy^2} = -R_i \tag{14}$$

Equation (14) is obtained from equation (7) by considering the following assumptions[13,43]:

- steady state $(\partial C_i / \partial t = 0)$,
- unidirectional flow (v_x is the non-zero velocity component, $v_y = v_z = 0$),
- axial diffusion negligible compared to axial convection $(D_i \partial^2 C_i / \partial x^2 \approx 0)$,
- fully developed flow (v_x does not change in axial direction),
- isothermal conditions (constant fluid properties).

Charles investigated the kinetics of photocatalytic degradation of salicylic acid in microreactors (1 2 mm channel diameter) with immobilized titanium dioxide as photocatalyst.[53] Their results high-light the importance of considering external mass-transfer limitations and the influence of axial dispersion in photoreactor modeling for determining the kinetic constants in microreactors.

Firstly, the kinetic parameters were determined by fitting the predicted results using the mass balance equation under the assumption of perfect plug-flow with the experimental results acquired under different residence times. In Figure 3.3(a), the experimental results are illustrated with symbols and the predictions are illustrated by lines. The continuous line was calculated with the rate constants (k_{app}, K_{app}) fitted at 4 mL/h flow rate, the dashed line with the parameters fitted at 10 mL/h and the dotted line with the parameters fitted at 20 mL/h. As it can be observed in Figure 3.3(a), the obtained apparent kinetic constants k_{app} and K_{app} vary with the flow rate and neither set of constants was able to match the degradation rate at all flow rates.

In a second case, Charles considered a model based on the plug-flow reactor assumption with convective mass transport and axial diffusion.[53] As it can be seen in Figure 3.3(b), the fitted kinetic parameters k (1.70×10^{-3} mmol L^{-1} s^{-1}) and K (24 L mmol^{-1}) were able to predict the degradation of salicylic acid at all flow rates. This shows that the obtained kinetic parameters are independent of the operating conditions. However, the rate constant k was found to be significantly dependent on the light intensity, as the radiation balance was not considered when developing the kinetic model.

Aillet *et al.* conducted kinetic investigations of a photochemical reaction in a spiral-shaped microreactor (0.5 mm inner diameter) irradiated by an ultraviolet/light-emitting diode array.[54] A T-photochromic system was studied which involves a reversible reaction between a closed form and an open form. The kinetic parameters of this reaction were determined using the modeling tools previously developed by the group.[13] The employed model used the mass balance as expressed in equation (14). In addition to the previously discussed study by Charles *et al.*,[53] the effect of the radiation intensity was considered in the reaction rate through the LVRPA.

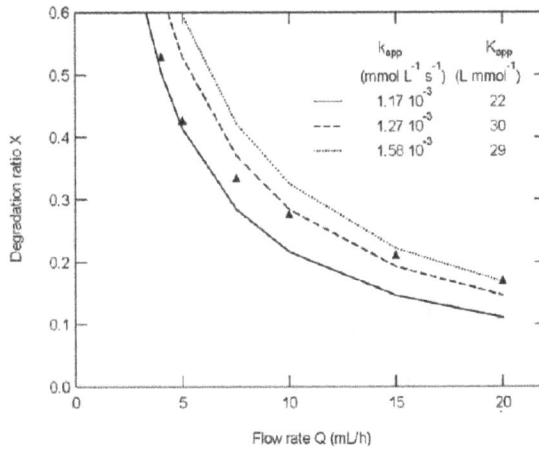

(a)

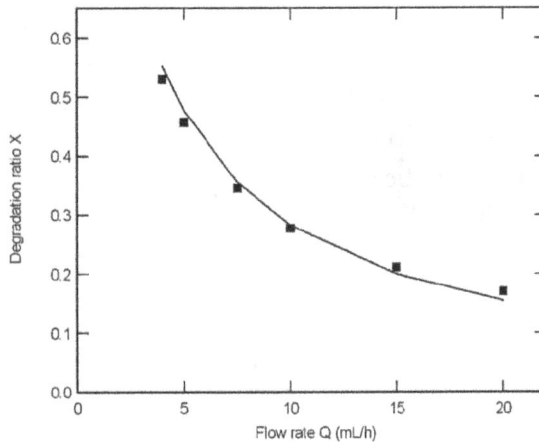

(b)

Fig. 3.3. The degradation ratio of salicylic acid (X) as a function of flow rate (Q). (a) Experimental degradation ratios (symbols) and calculated degradation ratios assuming a perfect plug-flow reactor (lines); the continuous line was calculated with the rate constants (k_{app}, K_{app}) fitted at 4 mL/h flow rate, the dashed line with the parameters fitted at 10 mL/h and the dotted line with the parameters fitted at 20 mL/h; (b) experimental degradation ratios (symbols) and calculated degradation ratios assuming a plug-flow reactor with mass transport limitations (line). Fitting parameters are $k = 1.70 \times 10^{-3}$ mmol L^{-1} s^{-1} and $K = 24$ L mmol^{-1}. All shown experimental results were acquired using a light source with $I_{incident} = 1.5$ mW/cm^2 and a microreactor with $d_{channel} = 2$ mm. (Reprinted with permission from Ref. 53. Copyright © 2011 Elsevier B.V.)

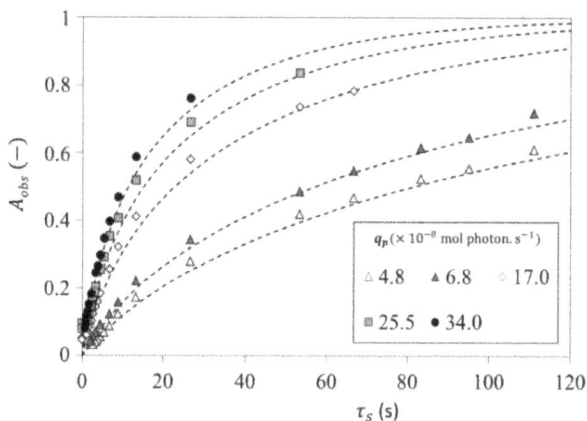

Fig. 3.4. The variation of the monitored absorbance (A_{obs}) at 610 nm with the residence time (τ_s) at different photon fluxes (q_p). Comparison between experimental (symbols) and predicted absorbance (dotted lines). (Reprinted with permission from Ref. 54. Copyright © 2016 WILEY-VCH Verlag GmbH & Co. KGaA, Weinheim.)

Therefore, the determined kinetic parameters were independent of the geometry, operating conditions and light source. This is proved in Figure 3.4 where the variation of absorbance, A_{obs} (the production of the open form product can be determined from the measured absorbance at 610 nm) with the residence time, τ_s, is illustrated. The symbols represent the experimental data and the dashed lines are the model predictions with the fitted kinetic parameters. As it can be observed, the determined kinetic parameters are able to predict the conversion variation with the residence time (flow rate) and with the photon flux, q_p.

3.5.2 *Mass balance in large-scale reactors*

The kinetic parameters estimated using a laboratory-scale reactor can be applied in the kinetic model, which in turn can then be included in the mass balance of a large-scale reactor.

Marugán *et al.* developed a model of a bench-scale, annular reactor for the photocatalytic disinfection of water (inactivation of Escherichia coli bacteria).[55] The reactor was operated under continuous flow with no perfect mixing.

The mass balance as expressed in equation (8) was applied to the annular reactor under the following assumptions:

- steady state $(\partial C_i/\partial t = 0)$,
- unidirectional axial flow (v_z is the non-zero velocity component, $v_r = v_\theta = 0$),
- azimuthal symmetry $(\partial/\partial\theta = 0)$,
- negligible axial diffusion compared to the convective flux in the axial direction $(D_i\partial^2 C_i/\partial z^2 \approx 0)$,
- fully developed flow (v_x does not change in axial direction, $v_x(r)$),
- isothermal conditions (constant fluid properties).

The resulting mass balance expression in cylindrical coordinates is given by

$$v_z\frac{\partial C_i}{\partial z} = D_i\left(\frac{1}{r}\frac{\partial}{\partial r}\left(r\frac{\partial C_i}{\partial r}\right)\right) + R_i \tag{15}$$

Figure 3.5 illustrates the variation of bacteria concentration expressed in colony forming units (CFU) with the catalyst loading after 2000 s of irradiation. Comparing the experimental results (symbols) with the predicted data (line), it can be observed that the

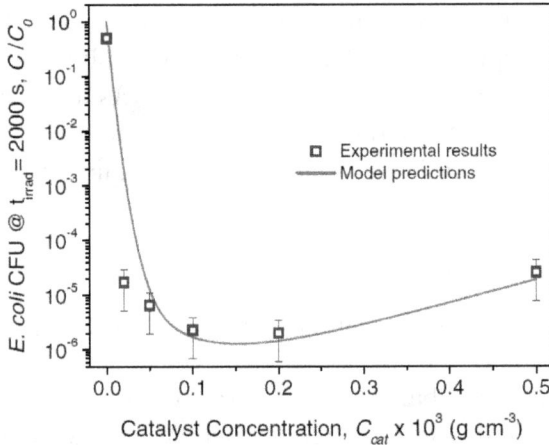

Fig. 3.5. Bacteria concentration after 2000 s as a function of catalyst concentration. Comparison between experimental results (symbols) and model predictions (line). (Reprinted with permission from Ref. 55. Copyright © 2012 Elsevier B.V.)

photoreactor model based on equation (15) was able to predict an optimal catalyst concentration in the range of $0.1-0.2 \times 10^{-3}$ gcm^{-3} catalyst concentration.

This optimum catalyst concentration range was not observed in the experiments which were carried out in the well-mixed laboratory-scale reactor for determining the kinetic parameters. The reduced activity observed in the presence of high catalyst loadings indicates the occurrence of mass transport limitations and should only appear in the non-perfectly mixed photoreactors systems.

3.6 Momentum Balance

3.6.1 *Conservation of momentum*

The velocity profile which has to be included in the mass balance can be determined from the momentum balance. The Cauchy momentum equation is a general statement of conservation of linear momentum. It provides the starting point for analyzing both Newtonian and non-Newtonian flows for either constant and variable density[56]:

$$\frac{\partial \rho \mathbf{v}}{\partial t} = -\nabla \cdot (\rho \mathbf{v} \mathbf{v}) - \nabla P - \nabla \cdot \boldsymbol{\tau} + \rho \mathbf{g} \qquad (16)$$

where ρ is the mass density, $\mathbf{v}$ is velocity vector, t is time, P is pressure, $\boldsymbol{\tau}$ is the viscous stress tensor and $\mathbf{g}$ is the gravitational acceleration vector.

$\partial \rho \mathbf{v}/\partial t$ represents the rate of increase of momentum per unit volume, $\nabla \cdot (\rho \mathbf{v} \mathbf{v})$ is the rate of momentum addition by convection per unit volume, ∇P and $\nabla \cdot \boldsymbol{\tau}$ represent the rate of momentum addition by molecular transport per unit volume and $\rho \mathbf{g}$ is the external force on fluid per unit volume.[57]

Considering Newtonian fluids with constant density (incompressible fluids), constant viscosity and inserting the Newtonian expression of τ in the previous equation results in the Navier–Stokes equations[56]:

$$\rho \left(\frac{\partial \mathbf{v}}{\partial t} + \mathbf{v} \cdot \nabla \mathbf{v} \right) = -\nabla P + \mu \nabla^2 \mathbf{v} + \rho \mathbf{g} \qquad (17)$$

Together with the continuity equation (assuming constant density):

$$\nabla \cdot \mathbf{v} = 0 \tag{18}$$

the Navier–Stokes equations allow to analyze the flow of liquids or gases at moderate velocities.[56]

The Navier–Stokes equations can be simplified depending on the considered situation, and in the following we will describe the simplifications for parallel-plate channel. This case is often encountered when describing flat-plate photoreactors. We start by considering the Navier–Stokes equations in component form in rectangular coordinates[56]:

$$\rho \left(\frac{\partial v_x}{\partial t} + v_x \frac{\partial v_x}{\partial x} + v_y \frac{\partial v_x}{\partial y} + v_z \frac{\partial v_x}{\partial z} \right)$$

$$= -\frac{\partial P}{\partial x} + \mu \left(\frac{\partial^2 v_x}{\partial x^2} + \frac{\partial^2 v_x}{\partial y^2} + \frac{\partial^2 v_x}{\partial z^2} \right) + \rho g_x \tag{19}$$

$$\rho \left(\frac{\partial v_y}{\partial t} + v_x \frac{\partial v_y}{\partial x} + v_y \frac{\partial v_y}{\partial y} + v_z \frac{\partial v_x}{\partial z} \right)$$

$$= -\frac{\partial P}{\partial y} + \mu \left(\frac{\partial^2 v_y}{\partial x^2} + \frac{\partial^2 v_y}{\partial y^2} + \frac{\partial^2 v_y}{\partial z^2} \right) + \rho g_y \tag{20}$$

$$\rho \left(\frac{\partial v_z}{\partial t} + v_x \frac{\partial v_z}{\partial x} + v_y \frac{\partial v_z}{\partial y} + v_z \frac{\partial v_z}{\partial z} \right)$$

$$= -\frac{\partial P}{\partial z} + \mu \left(\frac{\partial^2 v_z}{\partial x^2} + \frac{\partial^2 v_z}{\partial y^2} + \frac{\partial^2 v_z}{\partial z^2} \right) + \rho g_z \tag{21}$$

Next, the following assumptions are considered:

- steady state ($\partial/\partial t = 0$),
- unidirectional axial flow (v_x is the non-zero velocity component, $v_y = v_z = 0$; consequently, all terms in equations (20) and (21) vanish),
- fully developed flow ($\partial v_x/\partial x = 0$, then it also follows that $\partial^2 v_x/\partial x^2 = 0$),
- the z dimension is large enough that edge effects are negligible, which results in $v_x = v_x(y)$,

- no external force (e.g. the channel is placed horizontally, therefore $\rho g_x = 0$).

and the simplified equation can be written as

$$0 = -\frac{\partial P}{\partial x} + \mu \left(\frac{\partial^2 v_x}{\partial y^2}\right) \tag{22}$$

By integrating equation (22), the velocity profile in the parallel-plate channel can be obtained.

3.6.2 *Velocity profile in common photoreactor configurations*

The expressions for describing the velocity profile for the most encountered cases in photoreactor modeling are shown together with their applications reported in literature.

The velocity profile, $v_x(y)$, of a steady-state, fully developed laminar flow in parallel-plate channels can be expressed in rectangular coordinates as[47]:

$$v_x(y) = v_{max}\left[1 - \left(\frac{2y}{H} - 1\right)^2\right] \tag{23}$$

$$v_{max} = \frac{3}{2}\langle v_x \rangle \tag{24}$$

where v_{max} is the maximum fluid velocity in the channel, $\langle v_x \rangle$ is the mean fluid velocity in the channel, y is the rectangular coordinate and H is the height of the channel. The flow representation for the considered parallel-plate channel is illustrated in Figure 3.6(a).

Equation (23) was used under slightly different form in the modeling of flat-plate photoreactors[43,47] and photomicroreactors.[13,44]

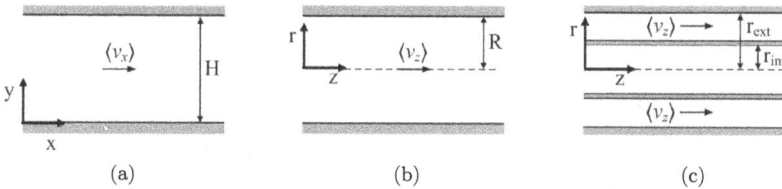

Fig. 3.6. Representation of the flow in (a) a parallel-plate channel; (b) a tubular channel; (c) an annular reactor.

The expression for the velocity profile $v_x(r)$ in a tubular channel for steady, fully developed laminar flow can be written in cylindrical coordinates as follows[56,57]:

$$v_x(r) = v_{max}\left[1 - \left(\frac{r}{R}\right)^2\right] \tag{25}$$

$$v_{max} = 2\langle v_z\rangle \tag{26}$$

where r is the cylindrical coordinate and R is the channel radius. The representation of the flow for the considered tubular channel is illustrated in Figure 3.6(b). The velocity profile in a circular tube is parabolic, as for the parallel-plate channel, but the maximum velocity in this case is $2\langle v\rangle$.[56] Equation (25) was used by Schechter and Wissler in the modeling of a tubular photoreactor.[58]

The velocity profile $v_z(r)$ through an annulus for steady-state, fully developed laminar flow can be written in cylindrical coordinates as follows[57]:

$$v_z(r) = v_{max}\frac{1 - \left(\frac{r}{r_{ext}}\right)^2 + \frac{1-\kappa^2}{\ln\left(\frac{1}{\kappa}\right)}\ln\left(\frac{r}{r_{ext}}\right)}{\frac{1-\kappa^4}{1-\kappa^2} - \frac{1-\kappa^2}{\ln\left(\frac{1}{\kappa}\right)}} \tag{27}$$

$$v_{max} = 2\langle v_z\rangle \tag{28}$$

where r is the cylindrical coordinate, r_{int} is the radius of the inner cylinder of the reactor, r_{ext} is the radius of the outer cylinder of the reactor and κ is the ratio r_{int}/r_{ext}. The flow representation for the considered annulus is shown in Figure 3.6(c). Equation (27) was used in modeling annular photoreactors with small form variations.[28,45,51,55,59]

3.7 Applications of Continuous-Flow Photoreactor Modeling

3.7.1 *Modeling for photoreactor scale-up*

Applications of the photoreactor scale-up methodology previously described are briefly presented in Table 3.2. Most of the applications are from the water and air treatment field, but the same scale-up

Table 3.2. Examples of scale-up to large continuous photoreactors using the methodology reported by Alfano and Cassano.[8]

| Application | Medium | Reactor configuration | | References |
		Laboratory scale	Large scale	
Oxidation of dilute aqueous HCOOH solutions	Liquid	Flat-plate	Annular	45
Degradation of perchloroethylene	Gas–solid	Flat-plate	Multiannular	16
Degradation of 4-chlorophenol	Liquid–solid (slurry)	Cylindrical	Flat-plate	5
Degradation of herbicides	Liquid–solid (slurry)	Annular	Falling film	7
Degradation of formaldehyde	Gas–solid	Flat-plate	Corrugated wall reactor	49
Inactivation of *Escherichia coli*	Liquid–solid (slurry)	Annular	Annular	55
Degradation of aqueous polyvinyl alcohol	Liquid	Annular	Annular	6

methodology may be applied also in other areas, such as synthetic chemistry. As it can be observed, the methodology was used with various reactor geometries in homogeneous (liquid) and heterogeneous phase (gas–solid, liquid–solid). The reported volumes of annular reactors range from $134\,cm^3$ at laboratory scale[7] and up to $6000\,cm^3$ at large scale.[6] The flat-plate reactors were scaled from $26\,cm^3$ at laboratory scale[49] to up to $734\,cm^3$ at large scale.[5]

The main purpose of this chapter was to demonstrate the use of modeling as a tool for the scale-up process, but it is also interesting to consider the use of modeling as a tool to investigate the behavior of photoreactors for reactor optimization, which we will do in the next section.

3.7.2 *Modeling for reactor optimization*

Similar to the previously discussed scale-up studies, modeling of the photoreactor behavior was mostly employed in the water and air treatment fields. In general, the classical photoreactor configurations are considered, e.g. annular, flat-plate. However, recently also an increasing interest in the modeling of continuous-flow microreactors is observed.

One of the main objectives of the optimization investigations are the identification of mass-transfer limitations (external and internal) and CFD is extensively used for predicting velocity fields, concentration gradients and the photoreaction rate (see Table 3.3).

Table 3.3. Example of applications of modeling for reactor optimization.

Application	Reactor	Investigation	References
Degradation of trichloroethylene	Annular	CFD modeling for predicting the performance of various photocatalytic reactor concepts	60
Degradation of an organic pollutant	Annular	Dependence of the degradation rate on dimensionless parameters	59
Degradation of dichloroacetic acid	Flat-plate	Mass transport limitations	61, 62
Degradation of ammonia and butyric acid	Annular	Influence of mass transfer on kinetics	63
Degradation of thrichloroethylene	Flat-plate	CFD modeling to study the velocity field and concentration gradient	64
Degradation of benzoic acid	Annular	CFD modeling to predict the degradation rate over different operating regimes	65
Degradation of shower water	Annular	CFD modeling for relating the degradation rate with operating regimes	32
Degradation of para-chlorobenzoic acid	Annular	CFD modeling for predicting the degradation rate	66
Degradation of salicylic acid	Microreactor	CFD modeling for investigating the influence of reactor geometry on photocatalytic efficiency	67

Table 3.3. (*Continued*)

Application	Reactor	Investigation	References
Non-catalyzed reaction	Microreactor	Diffusion limitations; guidelines for operating conditions required for avoiding diffusion limitations	13
Degradation of methylene blue	Flat-plate	CFD modeling for reactor optimization	68
UV disinfection	Baffled cylindrical	CFD modeling for photoreactor optimization	69

3.8 Conclusion

The purpose of this chapter was to present the most reported and used modeling strategy to assist in the scale-up of photoreactors. We emphasized the interconnection between the principal steps of the approach, such as kinetic modeling, together with the radiation, material and momentum balances. In addition, we highlighted the most common assumptions to simplify the equations, which allows model development and obtaining results with reasonable computational effort.

The discussed examples from literature were chosen to illustrate the effect of the simplifications on the predicted results. This modeling methodology has been validated in various applications and for various reactor configurations and can therefore be considered as a reliable tool to guide the scale-up of continuous-flow photoreactors. Furthermore, these validated models also represent an important means in predicting photoreactor behavior.

References

1. B. D. A. Hook, W. Dohle, P. R. Hirst, M. Pickworth, M. B. Berry and K. I. Booker-Milburn, *J. Org. Chem.* **70**, 7558 (2005).
2. T. Van Gerven, G. Mul, J. Moulijn and A. Stankiewicz, *Chem. Eng. Process.* **46**, 781 (2007).

3. Y. Su, N. J. W. Straathof, V. Hessel and T. Noël, *Chem. Eur. J.* **20**, 10562 (2014).
4. S. Elgue, T. Aillet, K. Loubiere, A. Conté, O. Dechy-Cabaret, L. Prat, C. R. Horn, O. Lobet and S. Vallon, *Chim. Oggi* **33**, 58 (2015).
5. M. L. Satuf, R. J. Brandi, A. E. Cassano and O. M. Alfano, *Catal. Today* **129**, 110 (2007).
6. S. Ghafoori, M. Mehrvar and P. K. Chan, *Chem. Eng. J.* **245**, 133 (2014).
7. G. Li Puma, B. Toepfer and A. Gora, *Catal. Today* **124**, 124 (2007).
8. O. M. Alfano and A. E. Cassano, in *Advances in Chemical Engineering*, Vol. 36, Eds. H. I. de Lasa and B. Serrano (Elsevier, 2009), pp. 229–287.
9. O. M. Alfano, R. L. Romero and A. E. Cassano, *Chem. Eng. Sci.* **41**, 421 (1986).
10. A. E. Cassano, C. A. Martin, R. J. Brandi and O. M. Alfano, *Ind. Eng. Chem. Res.* **34**, 2155 (1995).
11. T. Aillet, K. Loubiere, O. Dechy-Cabaret and L. Prat, *Chem. Eng. Process., Intensif.* **64**, 38 (2013).
12. A. E. Cassano and O. M. Alfano, *Catal. Today* **58**, 167 (2000).
13. T. Aillet, K. Loubière and L. Prat, *AIChE J.* **61**, 1284 (2015).
14. N. Serpone, A. Salinaro, *Pure Appl. Chem.* **71**, 303 (1999).
15. G. E. Imoberdorf, H. A. Irazoqui, A. E. Cassano and O. M. Alfano, *Ind. Eng. Chem. Res.* **44**, 6075 (2005).
16. G. E. Imoberdorf, H. A. Irazoqui, O. M. Alfano and A. E. Cassano, *Chem. Eng. Sci.* **62**, 793 (2007).
17. S. Ghafoori, M. Mehrvar and P. K. Chan, *Ind. Eng. Chem. Res.* **51**, 14980 (2012).
18. S. Ghafoori, M. Mehrvar and P. K. Chan, *Iran. Polym. J.* **21**, 869 (2012).
19. J. Marugán, R. van Grieken, A. E. Cassano and O. M. Alfano, *Appl. Catal. B.* **85**, 48 (2008).
20. V. Pareek, S. Chong, M. Tadé and A. A. Adesina, *Asia Pac. J. Chem. Eng.* **3**, 171 (2008).
21. G. Li Puma and A. Brucato, *Catal. Today* **122**, 78 (2007).
22. V. Pareek, M. P. Brungs and A. A. Adesina, *Ind. Eng. Chem. Res.* **42**, 26 (2003).
23. R. L. Romero, O. M. Alfano and A. E. Cassano, *Ind. Eng. Chem. Res.* **36**, 3094 (1997).
24. V. K. Pareek and A. A. Adesina, *AIChE J.* **50**, 1273 (2004).
25. M. A. Mueses, F. Machuca-Martinez, A. Hernández-Ramirez and G. Li Puma, *Chem. Eng. J.* **279**, 442 (2015).
26. A. L. L. Zazueta, H. Destaillats and G. Li Puma, *Chem. Eng. J.* **217**, 475 (2013).
27. T. Aillet, Conception et mise en oeuvre de reacteurs photochimiques intensifies, Ph.D. Thesis, Institut National Polytechnique de Toulouse (2015).
28. G. Li Puma, J. Khor and A. Brucato, *Environ. Sci. Technol.* **38**, 3737 (2004).
29. A. Brucato, A. E. Cassano, F. Grisafi, G. Montante, L. Rizzuti and G. Vella, *AIChE J.* **52**, 3882 (2006).

30. I. Grčić and G. Li Puma, *Environ. Sci. Technol.* **47**, 13702 (2013).
31. G. Sgalari, G. Camera-Roda and F. Santarelli, *Int. Commun. Heat Mass* **25**, 651 (1998).
32. Y. Boyjoo, M. Ang and V. Pareek, *Chem. Eng. Sci.* **111**, 266 (2014).
33. G. Spadoni, E. Bandini and F. Santarelli, *Chem. Eng. Sci.* **33**, 517 (1978).
34. M. Pasquali, F. Santarelli, J. F. Porter and P.-L. Yue, *AIChE J.* **42**, 532 (1996).
35. R. Changrani and G. B. Raupp, *AIChE J.* **45**, 1085 (1999).
36. G. E. Imoberdorf, F. Taghipour, M. Keshmiri and M. Mohseni, *Chem. Eng. Sci.* **63**, 4228 (2008).
37. J. Moreira, B. Serrano, A. Ortiz and H. de Lasa, *Ind. Eng. Chem. Res.* **49**, 10524 (2010).
38. M. Singh, I. Salvado-Estivill and G. Li Puma, *AIChE J.* **53**, 678 (2007).
39. A. Manassero, M. L. Satuf and O. M. Alfano, *Environ. Sci. Pollut. Res.* **22**, 926 (2015).
40. J. Pruvost, J. F. Cornet and J. Legrand, *Chem. Eng. Sci.* **63**, 3679 (2008).
41. M. Motegh, J. Cen, P. W. Appel, J. R. van Ommen and M. T. Kreutzer, *Chem. Eng. J.* **207–208**, 607 (2012).
42. M. A. Mueses, F. Machuca-Martinez and G. Li Puma, *Chem. Eng. J.* **215–216**, 937 (2013).
43. I. Salvadó-Estivill, A. Brucato and G. L. Puma, *Ind. Eng. Chem. Res.* **46**, 7489 (2007).
44. A. Visan, D. Rafieian, W. Ogieglo and R. G. H. Lammertink, *Appl. Catal. B* **150–151**, 93 (2014).
45. M. D. Labas, C. S. Zalazar, R. J. Brandi, C. A. Martín and A. E. Cassano, *Helv. Chim. Acta* **85**, 82 (2002).
46. A. Queffeulou, L. Geron, C. Archambeau, H. Le Gall, P.-M. Marquaire and O. Zahraa, *Ind. Eng. Chem. Res.* **49**, 6890 (2010).
47. R. J. Brandi, G. Rintoul, O. M. Alfano and A. E. Cassano, *Catal. Today* **76**, 161 (2002).
48. I. Salvadó-Estivill, D. M. Hargreaves and G. L. Puma, *Environ. Sci. Technol.* **41**, 2028 (2007).
49. C. Passalía, O. M. Alfano and R. J. Brandi, *J. Hazard. Mater.* **211–212**, 357 (2012).
50. G. Camera-Roda, V. Augugliaro, A. Cardillo, V. Loddo, L. Palmisano, F. Parrino and F. Santarelli, *Catal. Today* **259**, 87 (2016).
51. J. Marugán, R. van Grieken, A. E. Cassano and O. M. Alfano, *Catal. Today* **144**, 87 (2009).
52. R. Gorges, S. Meyer and G. Kreisel, *J. Photochem. Photobiol. A* **167**, 95 (2004).
53. G. Charles, T. Roques-Carmes, N. Becheikh, L. Falk, J.-M. Commenge and S. Corbel, *J. Photochem. Photobiol. A* **223**, 202 (2011).
54. T. Aillet, K. Loubière, O. Dechy-Cabaret and L. Prat, *Chem. Eng. Technol.* **39**, 115 (2016).

55. J. Marugán, R. van Grieken, C. Pablos, M. L. Satuf, A. E. Cassano and O. M. Alfano, *Chem. Eng. J.* **224**, 39 (2013).

56. W. M. Deen, *Analysis of Transport Phenomena*, 2nd Edn. (Oxford University Press, USA, 2013).

57. R. B. Bird, W. E. Stewart and E. N. Lightfoot, *Transport Phenomena*, 2nd Edn. (Wiley, 2002).

58. R. S. Schechter and E. H. Wissler, *Appl. Sci. Res.* **9**, 334 (1960).

59. G. Camera Roda and F. Santarelli, *Ind. Eng. Chem. Res.* **46**, 7637 (2007).

60. F. Taghipour and M. Mohseni, *AIChE J.* **51**, 3039 (2005).

61. M. D. L. M. Ballari, R. Brandi, O. Alfano and A. Cassano, *Chem. Eng. J.* **136**, 50 (2008).

62. M. D. L. M. Ballari, R. Brandi, O. Alfano and A. Cassano, *Chem. Eng. J.* **136**, 242 (2008).

63. B. Boulinguiez, A. Bouzaza, S. Merabet and D. Wolbert, *J. Photochem. Photobiol. A* **200**, 254 (2008).

64. A. Jarandehei, A. De Visscher, *AIChE J.* **55**, 312 (2009).

65. J. E. Duran, M. Mohseni and F. Taghipour, *AIChE J.* **57**, 1860 (2011).

66. M. Bagheri and M. Mohseni, *Chem. Eng. J.* **256**, 51 (2014).

67. S. Corbel, N. Becheikh, T. Roques-Carmes and O. Zahraa, *Chem. Eng. Res. Des.* **92**, 657 (2014).

68. V. Vaiano, O. Sacco, D. Pisano, D. Sannino and P. Ciambelli, *Chem. Eng. Sci.* **137**, 152 (2015).

69. H. Cao, B. Deng, J. Hong, J. Xue and F. Chang, *Chem. Eng. Technol.* **39**, 108 (2016).

Chapter 4

Photon Transport Phenomena: Radiation Absorption and Scattering Effects on Photoreactors

Orlando Mario Alfano,* María Lucila Satuf
and Agustina Manassero

*Instituto de Desarrollo Tecnológico
para la Industria Química
(Universidad Nacional del Litoral
and Consejo Nacional de Investigaciones
Científicas y Técnicas)
Colectora Ruta Nacional N° 168
3000 Santa Fe, Argentina
alfano@santafe-conicet.gov.ar

4.1 Introduction

Heterogeneous photocatalysis employing solid semiconductors has been applied to solve a wide variety of environmental problems involving the degradation of chemical pollutants and the inactivation of pathogen microorganisms, both in gas and liquid phase. Particularly, heterogeneous photocatalysis is able to transform refractory organics into readily biodegradable compounds, and eventually mineralized them into carbon dioxide and water.

All photocatalytic reactions are initiated by absorption of radiant energy by the catalyst, which promotes an electron from the valence band to a vacant conduction band, leaving a positive hole in the valence band. These charge carriers can then participate in reduction and oxidation reactions with chemical species adsorbed on the surface of the photocatalyst. The rate of the initial radiation-activated step is proportional to the absorbed energy through a property defined as the local volumetric rate of photon absorption (LVRPA).[1] The LVRPA represents the amount of photons that are absorbed per unit time and unit reaction volume. To evaluate the LVRPA, we must know the radiation intensity at each point inside the photocatalytic reactor. Intensity values can be obtained from the radiative transfer equation (RTE). To solve the RTE in slurry reactors, in which solid catalytic particles are suspended in aqueous solutions, the optical properties of the catalyst suspension are needed.

This chapter includes a derivation of the RTE together with the presentation of some methods that can be employed to solve this equation. A detailed description of the experimental method to measure and calculate the optical properties of the catalyst suspension is also given. Finally, a practical application showing the kinetic modeling of the photocatalytic degradation of an organic compound (the pharmaceutical clofibric acid (CA)) is presented. In this example, the optical properties experimentally measured are used in the LVRPA calculation.

4.2 Radiative Transfer Equation

4.2.1 *Radiation intensity*

Radiation Intensity is a fundamental quantity employed to describe the amount of radiation energy transmitted by a ray in any given direction per unit time. It can be defined as the radiation energy passing through a local area dA per unit time, per unit of the projected area, and per unit solid angle (Figure 4.1).

The projected area is formed by taking the area that the energy is passing through and projecting it normal to the direction of travel ($dA \cos \theta$). The elemental solid angle ($d\Omega$) is centered about the direction of travel ($\mathbf{\Omega}$) and has its origin at dA. The spectral (or monochromatic) intensity I_λ is the intensity per unit wavelength

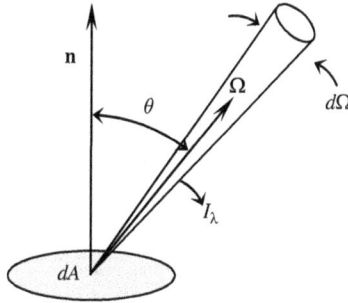

Fig. 4.1. Geometry for the definition of radiation intensity.

interval around a wavelength λ.

$$I_\lambda = \lim_{dA,\, d\Omega,\, d\lambda,\, dt \to 0} \left[\frac{dE_\lambda}{dA \cos\theta \, d\Omega \, d\lambda \, dt} \right] \tag{1}$$

4.2.2 *Derivation of the RTE*

The RTE describes the radiation intensity at any position along a path through an absorbing, emitting and scattering medium. In the specific case of heterogeneous photochemical reactors, radiation emission inside the reactor can be neglected because the process is generally performed at room temperature.[1]

Consider the spectral radiation intensity $I_\lambda(s, \mathbf{\Omega})$ that travels in an absorbing and scattering medium in the direction $\mathbf{\Omega}$ along a path (Figure 4.2).

As the radiation passes through ds, its intensity is reduced by absorption as

$$dI_{\lambda,a}(s, \mathbf{\Omega}) = -\kappa_\lambda(s) I_\lambda(s, \mathbf{\Omega}) ds \tag{2}$$

The quantity $dI_{\lambda,a}(s, \mathbf{\Omega})$ represents the decrease of the incident radiation $I_\lambda(s, \mathbf{\Omega})$ due to absorption. $\kappa_\lambda(s)$ is the spectral volumetric absorption coefficient, which represents the fraction of the incident radiation that is absorbed by the medium per unit length along the path s and has a unit of $(\text{length})^{-1}$.

When a photon interacts with one or more particles, it may undergo a loss (a change in direction) or a gain of energy. This phenomenon is called scattering. Scattering can be *elastic*, in which

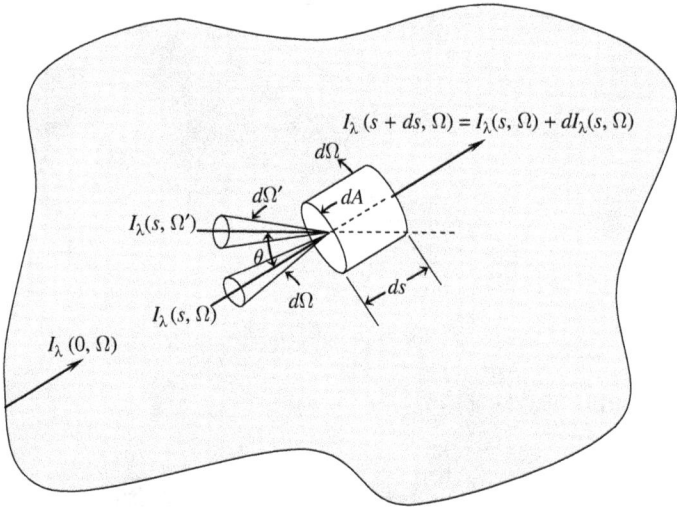

Fig. 4.2. Changes in radiation intensity in an absorbing and scattering medium.

the energy of the photon is unchanged, and *inelastic*, in which the energy is changed. Most scattering of importance in engineering is essentially elastic.[2] In turn, scattering can be *isotropic*, in which scattering into any direction is equally likely, and *anisotropic*, in which there is a distribution of scattering directions.

The attenuation of intensity due to scattering (out-scattering) can be written as

$$dI_{\lambda,o-s}(s, \mathbf{\Omega}) = -\sigma_\lambda(s)I_\lambda(s, \mathbf{\Omega})ds \qquad (3)$$

where $dI_{\lambda,o-s}(s, \mathbf{\Omega})$ indicates the decrease of the incident intensity due to scattering in all directions. The spectral volumetric scattering coefficient $\sigma_\lambda(s)$ represents the fraction of the incident radiation that is scattered by the medium in all directions per unit length along the path s and has a unit of (length)$^{-1}$.

The augmentation of intensity due to radiation incident from all directions that is scattered in the direction $\mathbf{\Omega}$ (in-scattering) is given by

$$dI_{\lambda,i-s}(s, \mathbf{\Omega}) = \frac{\sigma_\lambda(s)}{4\pi}ds \int_{\Omega'=0}^{4\pi} I_\lambda(s, \mathbf{\Omega}')p(\mathbf{\Omega}' \to \mathbf{\Omega})d\Omega' \qquad (4)$$

The angular distribution of the scattered intensity is described by the phase function $p(\mathbf{\Omega'} \to \mathbf{\Omega})$. The quantity $\frac{1}{4\pi}p(\mathbf{\Omega'} \to \mathbf{\Omega})d\Omega'$ represents the probability that the incident radiation at $\mathbf{\Omega'}$ will be scattered about the direction $\mathbf{\Omega}$.

Taking into account the loss of intensity due to absorption and out-scattering and the gain of intensity due to in-scattering in the considered direction, the change of intensity $I_\lambda(s, \mathbf{\Omega})$ as the radiation passes through distance ds results

$$dI_\lambda(s, \mathbf{\Omega}) = dI_{\lambda,a}(s, \mathbf{\Omega}) + dI_{\lambda,o-s}(s, \mathbf{\Omega}) + dI_{\lambda,i-s}(s, \mathbf{\Omega}) \qquad (5)$$

Introducing equations (2)–(4) into equation (5), and reordering, gives

$$\frac{dI_\lambda(s, \mathbf{\Omega})}{ds} = -\kappa_\lambda(s)I_\lambda(s, \mathbf{\Omega}) - \sigma_\lambda(s)I_\lambda(s, \mathbf{\Omega})$$

$$+ \frac{\sigma_\lambda(s)}{4\pi} \int_{\Omega'=0}^{4\pi} I_\lambda(s, \mathbf{\Omega'})p(\mathbf{\Omega'} \to \mathbf{\Omega})d\Omega' \qquad (6)$$

The two terms for attenuation by absorption and scattering can be combined, giving the equation of radiative transfer for absorbing and scattering media:

$$\frac{dI_\lambda(s, \mathbf{\Omega})}{ds} + \beta_\lambda(s)I_\lambda(s, \mathbf{\Omega}) = \frac{\sigma_\lambda(s)}{4\pi} \int_{\Omega'=0}^{4\pi} I_\lambda(s, \mathbf{\Omega'})p(\mathbf{\Omega'} \to \mathbf{\Omega}) \, d\Omega'$$

$$(7)$$

where $\beta_\lambda(s) = \kappa_\lambda(s) + \sigma_\lambda(s)$ is the spectral extinction coefficient.

The spectral albedo $\omega_\lambda(s)$ is defined as the ratio of the scattering coefficient to the extinction coefficient.

$$\omega_\lambda(s) \equiv \frac{\sigma_\lambda(s)}{\beta_\lambda(s)} = \frac{\sigma_\lambda(s)}{\kappa_\lambda(s) + \sigma_\lambda(s)} \qquad (8)$$

Introducing equation (8) into equation (7) gives

$$\frac{1}{\beta_\lambda(s)} \frac{dI_\lambda(s, \mathbf{\Omega})}{ds} + I_\lambda(s, \mathbf{\Omega}) = \frac{\omega_\lambda}{4\pi} \int_{\Omega'=0}^{4\pi} I_\lambda(s, \mathbf{\Omega'})p(\mathbf{\Omega'} \to \mathbf{\Omega})d\Omega' \quad (9)$$

Equation (9) is called "differential form" of the RTE, although it is, in fact, an integro-differential equation.

4.2.3 *Solution methods for the RTE*

In photocatalytic slurry reactors, where radiation absorption and scattering take place simultaneously, analytical solutions of the RTE are not possible and various numerical methods have been proposed. Deterministic methods that have been applied to solve the RTE include the P-N methods, Discrete Ordinate method, Finite Elements and Finite difference methods.[2] Among probabilistic methods, the most employed approach is the Monte Carlo (MC) method. This is a statistical method in which small quantities of energy are followed along their individual paths during radiative transfer. This method is relatively easy to formulate when dealing with complex problems that involve spectral effects and directional surfaces. In Section 4.4.2.3, we will present the solution of the RTE in a slurry reactor employing the Monte Carlo approach.

With the information of the local radiation intensities, the LVRPA can be calculated as

$$e^a(\boldsymbol{x}) = \int_\lambda \kappa_\lambda \int_{\Omega=4\pi} I_\lambda(\boldsymbol{x}, \boldsymbol{\Omega}) d\Omega d\lambda \qquad (10)$$

4.3 Extinction, Absorption and Scattering Coefficients

To solve the RTE in a heterogeneous photocatalytic reactor, three optical properties of the catalyst suspensions are required: the volumetric absorption coefficient κ_λ, the volumetric scattering coefficient σ_λ, and the phase function for scattering p. In what follows, a method to evaluate these coefficients is presented.[3] The method involves absorbance, diffuse reflectance and transmittance spectrophotometric measurements of TiO_2 suspensions, the evaluation of the radiation field in the spectrophotometer sample cell, and the application of a nonlinear optimization program to adjust model predictions to experimental data.

The aqueous suspensions for spectrophotometric measurements were prepared as follows: TiO_2 powder (P25, Evonik) was dried in an oven at 150°C for 12 h. Then, different amounts of catalyst were weighted, transferred to volumetric flasks and diluted to volume with ultrapure water. The samples were sonicated for 1 h and kept under magnetic stirring until measurement. The catalyst concentrations ranged from 0.2 to 2.0 g L^{-1}.

Spectrophotometric measurements of the suspensions were made by an Optronic OL series 750 spectroradiometer equipped with an OL 740-70 integrating sphere reflectance attachment. Rectangular quartz cells were employed.

4.3.1 *Absorbance measurements*

In a homogeneous medium, a spectrophotometric measurement of absorbance can be used to obtain the absorption coefficient. When working with heterogeneous samples, where absorption and scattering coexist, we can get the extinction coefficient ($\beta_\lambda = \kappa_\lambda + \sigma_\lambda$) from absorbance measurements under specially designed conditions to minimize contributions from in-scattering and simultaneously to minimize the collection by the instrument detector of the out-scattered rays originated at the sample cell. With this purpose, employing the same configuration as for standard absorbance measurements, a narrow vertical slit (20 mm high × 1.5 mm wide) was placed before the detector chamber of the spectrophotometer. This slit coincides exactly with the radiation beam coming out of an empty sample cell. At the same time, the sample cell was placed as far as possible from the detector position to facilitate the escape of out-scattered rays (in the forward direction) from the detector view angle.

Extinction coefficients from 275 to 405 nm were calculated from absorbance readings (ABS$_\lambda$), recorded as previously described, as $\beta_\lambda = 2.303 \, \text{ABS}_\lambda/L$, where L represents the cell path length. The specific extinction coefficients β_λ^* (per unit catalyst mass concentration C_m) were obtained by applying a standard linear regression to the plots of β_λ versus C_m.

4.3.2 *Diffuse transmittance and reflectance measurements*

For diffuse measurements, the integrating sphere attachment was employed. It is coated with poly-(tetrafluoroethylene) (PTFE) and has two openings in the wall for reflecting samples: the sample

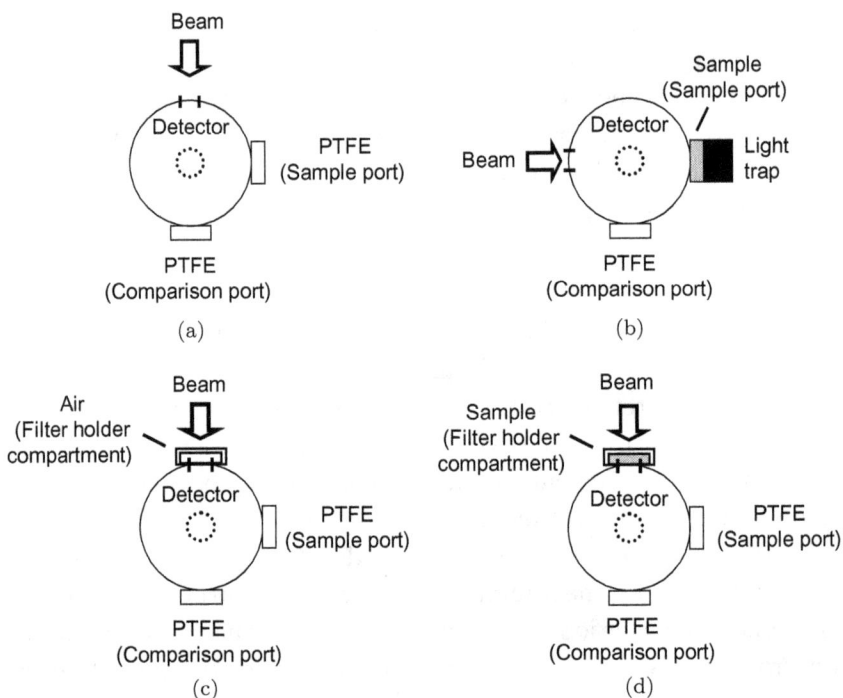

Fig. 4.3. Configuration of the integrating sphere for diffuse transmittance and reflectance measurements: (a) and (b) reflectance measurements; (c) and (d) transmittance measurements. (Reprinted with permission from Ref. 3. Copyright © 2005 American Chemical Society.)

port and the comparison port. The detector is positioned on a port mounted on the top of the integrating sphere. The OL 740-70 attachment also contains a filter holder compartment for transmittance measurements. The integrating sphere configurations for measurements are schematically shown in Figure 4.3. For diffuse reflectance measurements, a pressed PTFE reflectance standard was used as the reference in the comparison port. 100% reflectance reading was obtained placing another PTFE standard in the sample port (Figure 4.3(a)). To measure the reflectance of the sample, the quartz cell with the TiO_2 suspension was placed in the sample port with its back covered by a light trap that absorbs all transmitted radiation, keeping the PTFE standard in the comparison port (Figure 4.3(b)).

For diffuse transmittance measurements, PTFE reflectance standards
were placed in the comparison and sample ports. 100% transmittance
reading was obtained against air (Figure 4.3(c)). To measure the
transmittance of the sample, the cell with the catalyst suspension
was placed in the filter holder compartment (Figure 4.3(d)).

4.3.3 *Radiation field in the spectrophotometer sample cell*

The rectangular spectrophotometer cell, employed for reflectance and
transmittance measurements, can be represented as an infinite plane-
parallel medium with azimuthal symmetry (Figure 4.4).

Then, a one-dimensional, one-directional radiation transport
model was applied to solve the RTE in the cell; that is

$$\frac{\mu}{\beta_\lambda} \frac{\partial I_\lambda(x,\mu)}{\partial x} + I_\lambda(x,\mu) = \frac{\omega_\lambda}{2} \int_{\mu'=-1}^{1} I_\lambda(x,\mu')p(\mu,\mu')d\mu' \quad (11)$$

where x is the axial coordinate; μ is the direction cosine of the ray
for which the RTE is written ($\mu = \cos\,\theta$); and μ' is the cosine
of an arbitrary ray before scattering. Considering the quartz walls of
the cell as specularly reflecting surfaces, the boundary conditions of

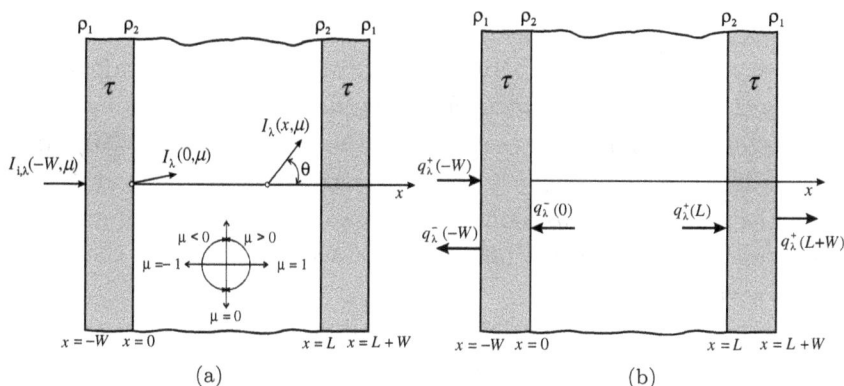

Fig. 4.4. Schematic representation of the spectrophotometer cell. (a) Coordinate system for the one-dimensional, one-directional radiation model. (b) Inlet
and outlet radiative fluxes. (Reprinted with permission from Ref. 3. Copyright ©
2005 American Chemical Society.)

equation (11) take the following form (Figure 4.4(a)):

$$I_\lambda(0, \mu) = \Gamma_{W2,\lambda} I_\lambda(0, -\mu) + Y_{W,\lambda} I_{i,\lambda}(-W, \mu) \quad (\mu > 0) \quad (12)$$

$$I_\lambda(L, \mu) = \Gamma_{W2,\lambda} I_\lambda(L, -\mu) \quad (\mu < 0) \quad (13)$$

where $Y_{W,\lambda}$ is the global wall transmission coefficient and $\Gamma_{W2,\lambda}$ the global wall reflection coefficient corresponding to the radiation that arrives from the internal side of the cell.[3]

4.3.4 *Phase function*

The choice of the phase function p represents an important step in any calculation where multiple scattering is involved. Actually, in a well-defined physical problem, the phase function is given, not chosen. Nevertheless, complicated functions lead to very time consuming computations. It is then a common practice to employ a phase function that preserves the main characteristics of the actual function and still renders the multiple-scattering computations manageable.[2]

The Henyey and Greenstein (HG) phase function $p_{\text{HG},\lambda}$ is characterized by

$$p_{\text{HG},\lambda}(\mu_0) = \frac{(1 - g_\lambda^2)}{(1 + g_\lambda^2 - 2g_\lambda\mu_0)^{3/2}} \quad (14)$$

where g_λ is the dimensionless asymmetry factor and μ_0 is the cosine of the scattering angle θ_0 between the directions of the incident and scattered rays. The HG function is then determined by a single free parameter (g_λ) that varies smoothly from isotropic $(g_\lambda = 0)$ to a narrow forward peak $(g_\lambda = 1)$ or to a narrow backward peak $(g_\lambda = -1)$. The knowledge of g_λ alone suffices to obtain solutions of multiple scattering problems with a high grade of accuracy, making the $p_{\text{HG},\lambda}$ ideal for calculations.[4]

To solve the RTE, the Discrete Ordinate Method[5] was applied. This method transforms the RTE into a set of algebraic equations that can be solved numerically. We are facing an inverse analysis of radiative transfer, where the optical properties ω_λ and g_λ will be determined from a set of measured radiation quantities. More details are given in Ref. 3.

4.3.5 *Parameters estimation*

In order to compare the model predictions with the experimental measurements of diffuse reflectance and transmittance, we should interpret the results in terms of net radiative fluxes. The spectral net radiative flux, $q_\lambda(x)$, for the one-dimensional, one-directional model with azimuthal symmetry, can be written as

$$q_\lambda(x) = 2\pi \int_{\mu=-1}^{1} I_\lambda(x,\mu)\mu\,d\mu \tag{15}$$

Accordingly, considering Figure 4.4(b), diffuse reflectance and transmittance values can be interpreted as the ratio between net radiative fluxes:

$$R_\lambda = \frac{q_\lambda^-(-W)}{q_\lambda^+(-W)} = \frac{2\pi \int_{\mu=-1}^{0} I_\lambda(-W,\mu)\mu\,d\mu}{2\pi \int_{\mu=0}^{1} I_{i,\lambda}(-W,\mu)\mu\,d\mu} \tag{16}$$

$$T_\lambda = \frac{q_\lambda^+(L+W)}{q_\lambda^+(-W)} = \frac{2\pi \int_{\mu=0}^{1} I_\lambda(L+W,\mu)\mu\,d\mu}{2\pi \int_{\mu=0}^{1} I_{i,\lambda}(-W,\mu)\mu\,d\mu} \tag{17}$$

Reflectance and transmittance model predictions are calculated with the values of I_λ obtained from the solution of the RTE. However, these values represent the radiation intensities at the inner walls of the cell ($x = 0$ and $x = L$). Therefore, to compare theoretical values with experimental results, one must account for the effect of the cell walls and obtain the corresponding values at $x = -W$ and $x = L + W$. The approach employed to calculate the intensities outside the sample cell is similar to the one applied for the boundary conditions of the RTE:

$$I_\lambda(-W,\mu) = \Gamma_{W1,\lambda}I_{i,\lambda}(-W,-\mu) + Y_{W,\lambda}I_\lambda(0,\mu) \quad (\mu < 0) \tag{18}$$
$$I_\lambda(L+W,\mu) = Y_{W,\lambda}I_\lambda(L,\mu) \quad (\mu > 0) \tag{19}$$

where $\Gamma_{W1,\lambda}$ represents the global wall reflection coefficient corresponding to the radiation that arrives from the external side of the cell.

A nonlinear multiparameter regression procedure (a modified Levenberg–Marquardt method) was applied to adjust theoretical values to experimental information. The optimization program renders the values of ω_λ and g_λ that minimize the differences between model predictions and experimental data. Then, for each concentration of TiO_2 and each wavelength, the volumetric scattering and absorption coefficients were obtained as $\sigma_\lambda = \beta_\lambda \omega_\lambda$ and $\kappa_\lambda = \beta_\lambda - \sigma_\lambda$. The corresponding specific coefficients ($\kappa_\lambda^* = \kappa_\lambda/C_m$, $\sigma_\lambda^* = \sigma_\lambda/C_m$), calculated per unit catalyst mass concentration, were obtained by applying a standard linear regression on the data κ_λ versus C_m and σ_λ versus C_m, respectively. The values of the dimensionless asymmetry factor, for each wavelength, were obtained by calculating the average value of g_λ for all the C_m tested in the experimental work.

Figure 4.5 shows that, for $\lambda < 320$ nm, κ_λ^* is larger than σ_λ^* (albedo < 0.5). For longer wavelengths, the specific absorption coefficient is always lower than the scattering coefficient. The asymmetry factor g_λ presents a smooth variation over the spectral range. It should be remarked that values of g_λ are positive over the whole range of wavelengths, ranging from 0.40 to 0.70, approximately. This behavior indicates a preferential forward direction of the scattered

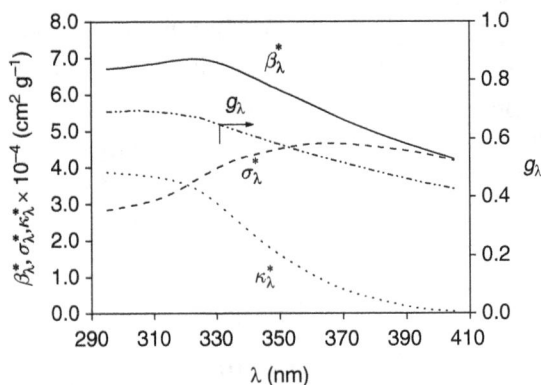

Fig. 4.5. Optical properties versus wavelength for Evonik P25 TiO_2 catalyst. Solid line: β_λ^*; broken line: σ_λ^*; dotted line: κ_λ^*; broken and dotted line, g_λ. (Reprinted with permission from Ref. 3. Copyright © 2005 American Chemical Society.)

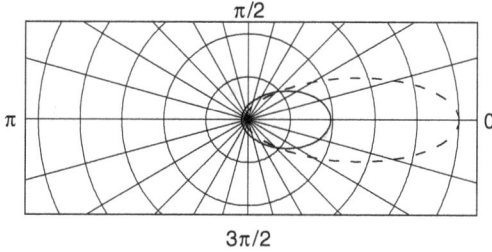

Fig. 4.6. HG phase function versus θ_0 for different values of g_λ. Solid line: $g_\lambda = 0.4$; broken line: $g_\lambda = 0.6$.

rays over titania particulate suspensions. Figure 4.6 shows the plots of the HG phase function versus θ_0, in polar coordinates, for two values of the asymmetry factor: 0.4 and 0.6.

4.4 Environmental Application: Kinetic Modeling in a Slurry Photocatalytic Reactor

In this section, the optical properties of the catalyst suspensions previously estimated are used to calculate the LVRPA in a slurry reactor. These results are then employed in the kinetic modeling of the degradation of an emerging pollutant: CA. CA is the active metabolite of clofibrate, a pharmaceutical widely employed as blood lipid regulator. The presence of pharmaceuticals in drinking water and natural water systems has increased public concern because of their potential adverse effects on human health and the aquatic environment. Although the concentrations of these compounds in water supplies are generally very low, their continuous input may lead to important long-term consequences in aquatic ecosystems.

4.4.1 *Experimental*

The reactor employed in the degradation experiments is cylindrical, made in glass, with two circular flat windows. It is illuminated through one of the circular windows, made of borosilicate ground glass. The schematic representation of the experimental setup for the photocatalytic reactions is shown in Figure 4.7. The reactor has an inner diameter of 5.0 cm and a length of 2.75 cm. It operates in

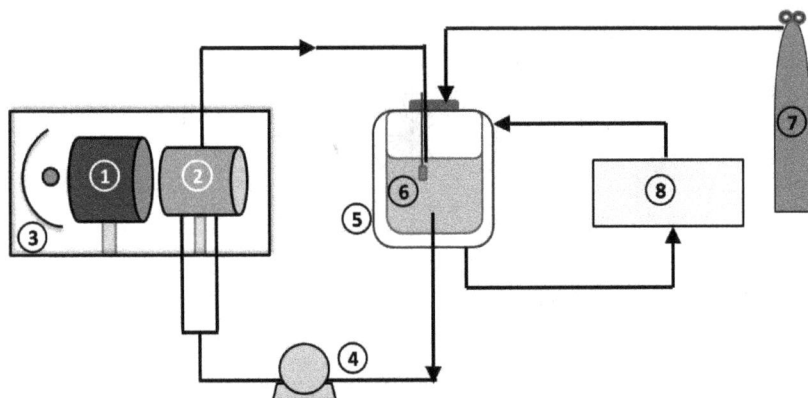

Fig. 4.7. Schematic representation of the experimental setup 1 — filter, 2 — reactor, 3 — lamp and reflector, 4 — pump, 5 — tank, 6 — thermometer, 7 — oxygen and 8 — thermostatic bath. (Reprinted with permission from Ref. 6. Copyright © 2014 Springer-Verlag, Berlin, Heidelberg.)

a closed recirculating circuit driven by a peristaltic pump (flow rate $1.5\,\text{L min}^{-1}$) with a storage tank of 1 L capacity. The tank is fitted with a device for withdrawal of liquid samples, a thermometer, a gas inlet for oxygen supply, and a water-circulating jacket to ensure isothermal conditions during the reaction time ($20°\text{C}$).

The light source is a halogenated mercury lamp (150-W Powerstar HQI from OSRAM), placed at the focal axis of a parabolic reflector. The lamp emits radiation in the UV–Vis range. In order to block the entrance of visible radiation to the reactor, a container with a solution of $CoSO_4$ was interposed between the reactor and the lamp. The wavelengths of the resulting radiation that arrives at the reactor window were comprised between 350 nm and 410 nm. Optical neutral filters were used to carry out experiments at different irradiation levels: 100%, 62%, and 30%. These filters attenuate the incident radiation without altering the spectral distribution of the lamp emission.

Aqueous suspensions of TiO_2 (Aeroxide P25, Evonik) and the pollutant were irradiated during 360 min and samples were taken every 60 min. The initial concentration of CA in all experiments was $9.30 \times 10^{-8}\,\text{mol cm}^{-3}$.

The concentration of CA and its main photocatalytic intermediates, 4-chlorophenol (4-CP) and benzoquinone (BQ), were measured by HPLC with a UV detector and an RP C-18 column.

Preliminary assays were carried out to evaluate (i) the direct photolysis of CA, (ii) the time necessary to reach the adsorption equilibrium between CA and TiO$_2$, and (iii) the optimum pH to carry out the reaction. These assays showed that the pollutant does not undergo photolysis under the conditions of the experiments. In addition, it was found that the adsorption equilibrium between CA and the catalyst was achieved after 1 h of recirculation in the reactor. Finally, the natural pH of the suspension (pH 5) was the most favorable to carry out the experiments.

A first set of experiments was carried out with a level of irradiation of 100%, employing different catalyst concentrations: 0.1, 0.25, 0.5 and 1.0 g L^{-1}. A second set of experiments was performed with a TiO$_2$ concentration of 0.5 g L^{-1} and two different levels of irradiation: 62% and 30%. More detailed information about the experimental procedure can be found in Ref. 6.

4.4.2 *Theoretical*

4.4.2.1 *Mass balances*

The theoretical evolution of CA, 4-CP and BQ concentrations was obtained from the mass balances for each compound. The assumptions made to obtain these balances were the following: the conversion per pass in the reactor is differential, the system is well-mixed, mass-transfer limitations are negligible, direct photolysis is insignificant, and chemical reactions occur only at the solid–liquid interface among adsorbed molecules. Thus, the general form of the mass balance for a generic compound i (i = CA, 4-CP, BQ) can be written as[7]:

$$\varepsilon_L \frac{dC_i(t)}{dt}\bigg|_{Tk} = \frac{V_R}{V_T} a_\nu \nu_i \langle r(x,t) \rangle_{A_R} \tag{20}$$

where ε_L is the liquid hold-up ($\varepsilon_L \cong 1$), C_i is the molar concentration of i (mol cm^{-3}), t denotes to the reaction time (s), Tk refers to the tank, V_R and V_T are the reactor and total system volumes, respectively, a_ν is the catalytic surface area per unit suspension volume, ν_i is the stoichiometric coefficient, and $\langle r(x,t) \rangle_{A_R}$ is the superficial reaction rate averaged over the catalytic reaction area A_R. The value of a_ν is computed from the product of the catalyst specific surface area (S_g) and the catalyst concentration (C_m).

The reaction pathway for the photocatalytic degradation of CA[8] involves the formation of two main reaction intermediates from CA: 4-CP and BQ. This pathway also considers the formation of BQ from 4-CP. Next, the degradation of these two main intermediaries leads to the formation of secondary organic intermediates which could be eventually mineralized to CO_2, HCl, and H_2O.

The resulting mass balances for CA, 4-CP and BQ with the corresponding initial conditions are:

$$\varepsilon_L \frac{dC_{CA}(t)}{dt}\bigg|_{Tk} = -\frac{V_R}{V_T} a_\nu \{ \langle r_{CA,1}(x,t) \rangle_{A_R} + \langle r_{CA,2}(x,t) \rangle_{A_R} \}$$

$$C_{CA}(t=0) = C_{CA,0} \tag{21}$$

$$\varepsilon_L \frac{dC_{4-CP}(t)}{dt}\bigg|_{Tk} = \frac{V_R}{V_T} a_\nu \{ \langle r_{CA,1}(x,t) \rangle_{A_R} - \langle r_{4-CP,1}(x,t) \rangle_{A_R}$$

$$- \langle r_{4-CP,2}(x,t) \rangle_{A_R} \}$$

$$C_{4-CP}(t=0) = 0 \tag{22}$$

$$\varepsilon_L \frac{dC_{BQ}(t)}{dt}\bigg|_{Tk} = \frac{V_R}{V_T} a_\nu \{ \langle r_{CA,2}(x,t) \rangle_{A_R} + \langle r_{4-CP,2}(x,t) \rangle_{A_R}$$

$$- \langle r_{BQ}(x,t) \rangle_{A_R} \}$$

$$C_{BQ}(t=0) = 0 \tag{23}$$

where $r_{CA,1}$ and $r_{CA,2}$ represent the degradation rates of CA to give the intermediate 4-CP and BQ, respectively. Furthermore, $r_{4-CP,1}$ and $r_{4-CP,2}$ refer to the rate of degradation of 4-CP to generate secondary organic intermediates and BQ, respectively. Finally, r_{BQ} represents the degradation rate of BQ.

To solve the mass balances, it is necessary to derive a kinetic model to obtain the reaction rate expressions for CA and its intermediates.

4.4.2.2 *Kinetic model*

The reaction scheme proposed for the photocatalytic degradation of CA is shown in Table 4.1.[9,10]

The reaction rate expressions were obtained based on the above scheme and taking into account the following assumptions: (i) photocatalytic reactions occur at the surface of the catalyst particles among adsorbed molecules,[11] (ii) dynamic equilibrium is achieved

Table 4.1. Reaction scheme for the photocatalytic degradation of CA. (Reprinted with permission from Ref. 6. Copyright © 2014 by Springer-Verlag, Berlin, Heidelberg.)

Stage	Reaction	Rate
Activation	$TiO_2 + h\nu \rightarrow e^- + h^+$	r_{gs}
Recombination	$e^- + h^+ \rightarrow heat$	$k_2[e^-][h^+]$
Electron trapping	$e^- + O_{2,ads} \rightarrow \bullet O_2^-$	$k_3[e^-][O_{2,ads}]$
Hole trapping	$h^+ + H_2O_{ads} \rightarrow \bullet OH + H^+$ $h^+ + OH^-_{ads} \rightarrow \bullet OH$	$k_4[h^+][H_2O_{ads}]$
Hydroxyl attack	$CA_{ads} + \bullet OH \rightarrow 4-CP$ $CA_{ads} + \bullet OH \rightarrow BQ$ $4-CP_{ads} + \bullet OH \rightarrow X_i$ $4-CP_{ads} + \bullet OH \rightarrow BQ$ $BQ_{ads} + \bullet OH \rightarrow X_j$ $Y_{i,ads} + \bullet OH \rightarrow Y_m$	$k_5CA_{ads}][\bullet OH]$ $k_6[CA_{ads}][\bullet OH]$ $k_7[4-CP_{ads}][\bullet OH]$ $k_8[4-CP_{ads}][\bullet OH]$ $k_9[BQ_{ads}][\bullet OH]$ $k_1'[Y_{i,ads}][\bullet OH]$
Adsorption	$site_{O_2} + O_2 \leftrightarrow O_{2,ads}$ $site_{H_2O} + H_2O \leftrightarrow H_2O_{ads}$ $site_{H_2O} + H_2O \leftrightarrow OH^-_{ads} + H^+$ $site_{CA} + CA \leftrightarrow CA_{ads}$ $site_{CA} + 4-CP \leftrightarrow 4-CP_{ads}$ $site_{CA} + BQ \leftrightarrow BQ_{ads}$ $site_{Y_l} + Y_l \leftrightarrow Y_{l,ads}$	

between the bulk and the adsorbed concentrations of H_2O, O_2, inorganic species and organic compounds,[10,12] (iii) molecular oxygen and organic compounds are adsorbed on different sites of the catalyst particle,[9,13] (iv) competitive adsorption between CA and its main reaction intermediates is postulated, (v) the attack by the hydroxyl radical is the main degradation mechanism for CA and its intermediates,[14,15] (vi) O_2 concentration is constant and in excess with respect to the stoichiometric demand,[7] (vii) the concentration of water molecules and hydroxyl ions on the catalytic surface remains constant,[7] (viii) the superficial concentration of total absorption sites for CA, per unit area of catalyst, can be considered constant, and (ix) the rate of electron–hole generation is defined as $r_{gs}(x) = \frac{\bar{\phi}}{a_v} \int_\lambda e_\lambda^a(x) d\lambda$, where $\bar{\phi}$ is the primary quantum yield averaged over the wavelength range.[16] From these assumptions, the following reaction rate expressions were obtained:

$$r_{\text{CA}}(x,t) = \frac{(\alpha_{2,1} + \alpha_{2,2})C_{\text{CA}}(x,t)}{1 + \alpha_3 C_{\text{CA}}(x,t) + \alpha_1' C_{4-\text{CP}}(x,t) + \alpha_2' C_{\text{BQ}}(x,t)}$$

$$\times \left(-1 + \sqrt{1 + \frac{\alpha_1}{a_v} e^a(x)} \right) \qquad (24)$$

$$r_{4-\text{CP}}(x,t) = \frac{(\alpha_{4,1} + \alpha_{4,2})C_{4-\text{CP}}(x,t)}{1 + \alpha_3 C_{\text{CA}}(x,t) + \alpha_1' C_{4-\text{CP}}(x,t) + \alpha_2' C_{\text{BQ}}(x,t)}$$

$$\times \left(-1 + \sqrt{1 + \frac{\alpha_1}{a_v} e^a(x)} \right) \qquad (25)$$

$$r_{\text{BQ}}(x,t) = \frac{\alpha_5 C_{\text{BQ}}(x,t)}{1 + \alpha_3 C_{\text{CA}}(x,t) + \alpha_1' C_{4-\text{CP}}(x,t) + \alpha_2' C_{\text{BQ}}(x,t)}$$

$$\times \left(-1 + \sqrt{1 + \frac{\alpha_1}{a_v} e^a(x)} \right) \qquad (26)$$

where α_i are intrinsic kinetic parameters. A detailed derivation of these equations can be found in Ref. 6.

4.4.2.3 *LVRPA*

The radiation absorbed in the reactor was computed by the Monte Carlo method. This method consists of tracking the trajectory of a great number of photons and computing the locations where they are absorbed. Then, it is possible to calculate the spatial distribution of the rate of photon absorption inside the photocatalytic reactor, $e^a(x)$. The direction, the length of flight and the fate of the photons (i.e. absorption, scattering or loss) are calculated by using numbers randomly generated (R_i) between 0 and 1. Taking into account that absorption and scattering phenomena depend on the photon wavelength, the wavelength range of the radiation that enters the reactor was discretized into eight wavelengths. Also, the reactor length was divided into small cells to store the position of the absorbed photons.

The events that were taken into account in the photon tracking are briefly described below:

(i) Photon direction at the reactor window (θ): Due to the ground glass window, radiation emitted from the lamp that arrives at the inner side of the reactor window is diffuse. As a result, the same probability is assigned to all the directions.[17] The angle θ between the photon direction and the x-axis is given by

$$\theta = a\sin(2R_1 - 1) \tag{27}$$

(ii) Length of the photon flight (l): The photon travels a distance l in the reacting medium without interaction. Then, the value of l is given by[18]:

$$l = -\frac{1}{\beta_\lambda}\ln(1 - R_2) \tag{28}$$

where β_λ is the spectral extinction coefficient of the catalyst suspension (see Section 4.3.1). Then, the new location of the photon after traveling the distance l in the direction of θ, is determined by

$$x_{\text{new}} = x_{\text{old}} + e_x l \tag{29}$$

where x_{old} refers to the previous location of the photon inside the reactor, x_{new} refers to the new location, and e_x is the direction cosine.

(iii) Fate of the photon: If the new position of the photon lies outside the reactor, the photon is considered lost and the process is re-initiated. If the photon remains inside the reactor after traveling a distance l, it strikes a catalyst particle and two possibilities can take place: it can be absorbed or scattered. The probability to occur one event or the other is related to the albedo (see equation (8)). At higher ω_λ, higher the probability of scattering.[17] Therefore, if:

$$1 - \omega_\lambda \geq R_3 \qquad (30)$$

the photon is absorbed, stored in the corresponding cell and the trajectory ends. Otherwise, if the photon is scattered, a new direction is determined by the Henyey and Greenstein phase function.[2] The cosine of the angle that determines the new direction of the photon is given by[19,20]:

$$\cos\theta = \frac{1}{2g_\lambda}\left[1 + g_\lambda^2 - \left(\frac{1-g_\lambda^2}{1+g_\lambda(2R_4-1)}\right)^2\right] \qquad (31)$$

Then, a new value of l is estimated as described in step (ii), and the sequence is repeated until the photon is either absorbed or lost. Finally, the LVRPA is calculated taking into account the number of photons of wavelength λ absorbed in each cell ($n_{ph\lambda,abs}$) and the total number of photons considered in the simulation ($n_{ph,T}$) according to:

$$e^a(x) = \sum_{\lambda=350\,\text{nm}}^{\lambda=410\,\text{nm}} \frac{q_{w\lambda}n_{ph\lambda,abs}(x)}{n_{ph,T}\Delta x} \qquad (32)$$

where $q_{w\lambda}$ is the radiation flux of wavelength λ incident at the reactor window, experimentally measured by ferrioxalate actinometry,[21] and Δx is the length of the position cell.

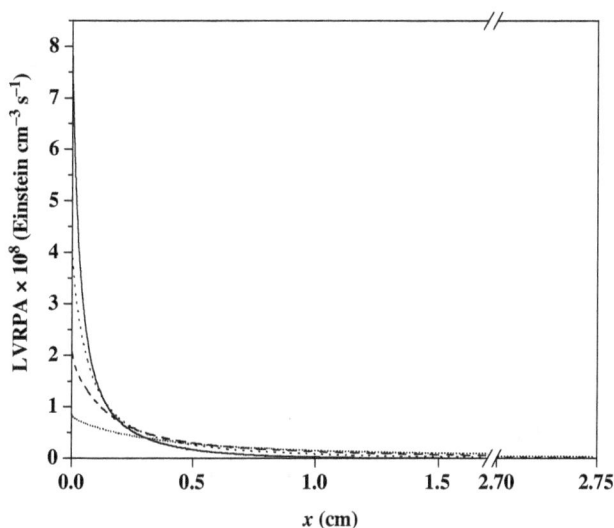

Fig. 4.8. LVRPA profiles for different catalyst concentrations. $0.1\,\mathrm{g\,L^{-1}}$ (short dot line), $0.25\,\mathrm{g\,L^{-1}}$ (dash line), $0.5\,\mathrm{g\,L^{-1}}$ (dot line) and $1\,\mathrm{g\,L^{-1}}$ (solid line). (Reprinted with permission from Ref. 6. Copyright © 2014 Springer-Verlag, Berlin, Heidelberg.)

The LVRPA profiles obtained with MC simulations for different catalyst concentrations are depicted in Figure 4.8. It should be noted that almost all the absorbed radiation was achieved in the space comprised between the irradiated window and $x = 1.0\,\mathrm{cm}$. The LVRPA profiles for 30% and 62% of irradiation level were also calculated. They can be found in Ref. 6.

Dependence of reaction kinetics with respect to the photon absorption rate

A nonlinear parameter estimator, the Levenberg–Marquardt algorithm, was employed to compare the experimental concentrations of CA, 4-CP and BQ and the corresponding concentrations predicted by the model. Based on these results, the terms $\alpha_3 C_{\mathrm{CA}}$, $\alpha'_1 C_{\text{4-CP}}$ and $\alpha'_2 C_{\mathrm{BQ}}$ were neglected because the values obtained were much lower than 1. Therefore, a new estimation was carried out with the following kinetic expressions:

$$r_{CA}(x,t) = (\alpha_{2,1} + \alpha_{2,2})C_{CA}(x,t)\left(-1 + \sqrt{1 + \frac{\alpha_1}{a_v}e^a(x)}\right)$$

$$\tag{33}$$

$$r_{4\text{-}CP}(x,t) = (\alpha_{4,1} + \alpha_{4,2})C_{4\text{-}CP}(x,t)\left(-1 + \sqrt{1 + \frac{\alpha_1}{a_v}e^a(x)}\right)$$

$$\tag{34}$$

$$r_{BQ}(x,t) = \alpha_5 C_{BQ}(x,t)\left(-1 + \sqrt{1 + \frac{\alpha_1}{a_v}e^a(x)}\right) \tag{35}$$

The values of the six parameters obtained with the optimization procedure are shown at the top of Table 4.2.

At this point, it is worth noting that the kinetics of photocatalytic reactions in slurry reactors is always a function of the photon absorption by the TiO_2 suspension. This dependence is given by the term $\left(-1 + \sqrt{1 + \frac{\alpha_1}{a_v}e^a(x)}\right)$ in equations (33)–(35). This term can take two limiting values: (i) for high radiation levels and, therefore, high values of e^a, the reaction rate is proportional to the square root

Table 4.2. Estimated parameters for the 6- and 5-parameters kinetic models.

Parameter	Value	Units
6-parameters kinetic model		
α_1	6.07×10^{11}	s cm^2 Einstein^{-1}
$\alpha_{2,1}$	5.83×10^{-6}	cm s^{-1}
$\alpha_{2,2}$	6.10×10^{-7}	cm s^{-1}
$\alpha_{4,1}$	1.41×10^{-6}	cm s^{-1}
$\alpha_{4,2}$	7.97×10^{-6}	cm s^{-1}
α_5	4.77×10^{-4}	cm s^{-1}
5-parameters kinetic model		
$\alpha'_{2,1}$	3.497	s$^{-0.5}$ cm^2 Einstein$^{-0.5}$
$\alpha'_{2,2}$	0.451	s$^{-0.5}$ cm^2 Einstein$^{-0.5}$
$\alpha'_{4,1}$	0.750	s$^{-0.5}$ cm^2 Einstein$^{-0.5}$
$\alpha'_{4,2}$	4.618	s$^{-0.5}$ cm^2 Einstein$^{-0.5}$
α'_5	3.71×10^2	s$^{-0.5}$ cm^2 Einstein$^{-0.5}$

of the photon absorption rate, and (ii) for low radiation levels there is a linear dependence of the reaction rate with the photon absorption rate.[16] For the experimental conditions adopted in this work, the first case was corroborated. Thus, the final kinetic expressions take the following form:

$$r_{\mathrm{CA}}(x,t) = (\alpha'_{2,1} + \alpha'_{2,2})C_{\mathrm{CA}}(x,t)\sqrt{e^a(x)} \tag{36}$$

$$r_{\mathrm{4\text{-}CP}}(x,t) = (\alpha'_{4,1} + \alpha'_{4,2})C_{\mathrm{4\text{-}CP}}(x,t)\sqrt{e^a(x)} \tag{37}$$

$$r_{\mathrm{BQ}}(x,t) = \alpha'_5 C_{\mathrm{BQ}}(x,t)\sqrt{e^a(x)} \tag{38}$$

where $\alpha'_{2,1} = \alpha_{2,1}\sqrt{\alpha_1/a_v}$, $\alpha'_{2,2} = \alpha_{2,2}\sqrt{\alpha_1/a_v}$, $\alpha'_{4,1} = \alpha_{4,1}\sqrt{\alpha_1/a_v}$, $\alpha'_{4,2} = \alpha_{4,2}\sqrt{\alpha_1/a_v}$ and $\alpha'_5 = \alpha_5\sqrt{\alpha_1/a_v}$.

Consequently, under this limiting case, the number of parameters to be estimated decreases from 6 to 5. At the bottom of the Table 4.2, the new values of the five parameters are presented.

It is worth mentioning that employing both kinetic models, good agreement between the experimental data and the predicted results was obtained, pointing out that the simplification proposed for high radiation levels is valid.

4.4.3 *Results and discussion*

Figure 4.9 depicts the experimental concentrations of CA, 4-CP and BQ, and the concentrations estimated by the 5-parameters model for different catalyst concentrations. It should be noted that the photocatalytic reaction rate increased when increasing the TiO$_2$ concentration from 0.25 to 1.0 g L^{-1}. More details about the effects of the catalyst loading on the reaction rates can be found in Ref. 6.

The root mean square error (RMSE) of predictions was estimated according to the following equation:

$$\mathrm{RMSE}\,\% = \sqrt{\frac{1}{N}\sum_{i=1}^{N}\left(\frac{C_{\mathrm{exp},i} - C_i}{C_{\mathrm{exp},i}}\right)^2} \times 100 \tag{39}$$

where $C_{i,\mathrm{exp}}$ and C_i are the experimental and theoretical concentrations of compound i, respectively. Also, N represents the total

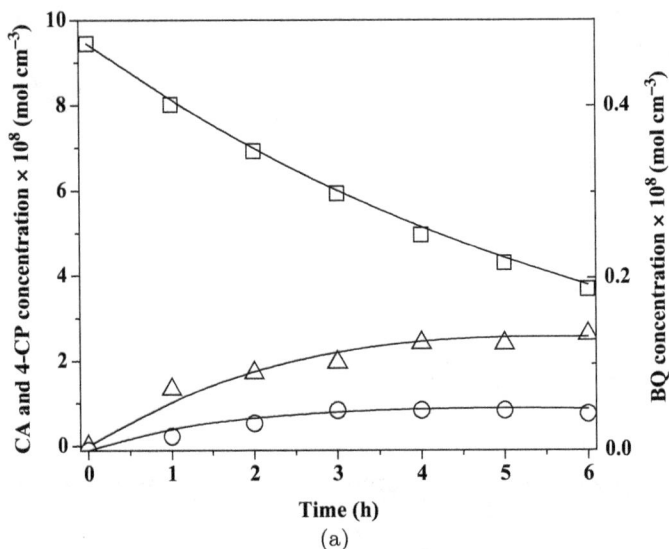

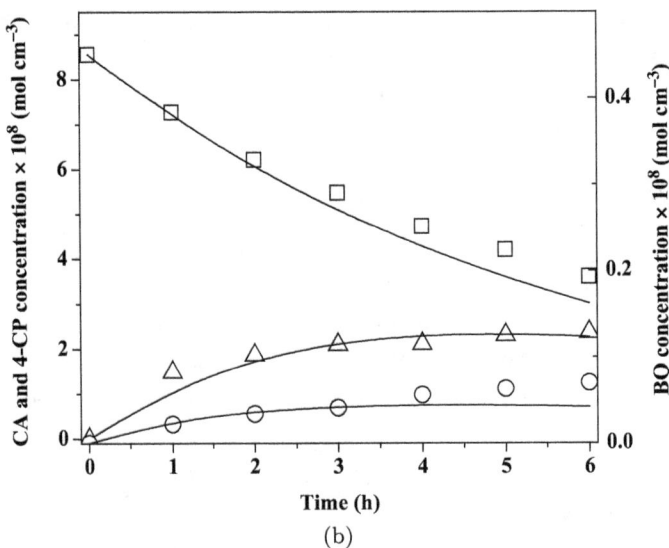

Fig. 4.9. Experimental and predicted concentrations of CA, 4-CP and BQ versus time for 100% of irradiation. Experimental data: ($\square$) CA; ($\triangle$) 4-CP and ($\bigcirc$) BQ. Model results: solid lines. (a) $C_m = 0.25$ g L^{-1} and (b) $C_m = 1.0$ g L^{-1}. (Reprinted with permission from Ref. 6. Copyright © 2014 Springer-Verlag, Berlin, Heidelberg.)

number of concentrations evaluated. The RMSE of predictions corresponding to the main pollutant was 5.9%. Considering the three organic compounds, the RMSE was 16.6%. From these results, it can be concluded that the model can simulate the evolution of CA, 4-CP and BQ in a photocatalytic slurry reactor under different catalyst concentrations and irradiation levels.

4.5 Conclusion

Reliable kinetic information is essential to design efficient photocatalytic devices for water detoxification, which should tend to achieve maximum interaction between radiation, catalyst and pollutant molecules. In this chapter, a methodology to model and evaluate the absorption of radiation in photocatalytic slurry reactors is presented, which includes the measurement of the optical properties of the catalyst suspension and the solution of the equation of radiative transfer in the reactor. Afterwards, this information was explicitly included in kinetic expressions that represent the degradation of organic pollutants. Finally, model results were adjusted to experimental data to estimate kinetic parameters. The resulting parameters are independent of the irradiation conditions and concentrations of catalyst, within the limits of the experimental conditions tested, and can be applied to design, scale-up, or optimize slurry reactors of different configurations.

Acknowledgments

The authors are grateful to Universidad Nacional del Litoral (UNL), Consejo Nacional de Investigaciones Científicas y Técnicas (CONICET), and Agencia Nacional de Promoción Científica y Tecnológica (ANPCyT) for financial support. We also thank Antonio C. Negro for his valuable help during the experimental work.

References

1. A. E. Cassano, C. A. Martín, R. J. Brandi and O. M. Alfano, *Ind. Eng. Chem. Res.* **34**, 2155 (1995).

2. R. Siegel and J. R. Howell, *Thermal Radiation Heat Transfer* (Hemisphere Publishing Corp., Bristol, 2002).
3. M. L. Satuf, R. J. Brandi, A. E. Cassano and O. M. Alfano, *Ind. Eng. Chem. Res.* **44**, 6643 (2005).
4. H. C. Van de Hulst, *Multiple Light Scattering* (Academic Press, New York, 1980).
5. J. J. Duderstadt and W. R. Martin, *Transport Theory* (Wiley, New York, 1979).
6. A. Manassero, M. L. Satuf and O. M. Alfano, *Environ. Sci. Pollut. Res.* **22**, 926 (2015).
7. M. L. Satuf, R. J. Brandi, A. E. Cassano and O. M. Alfano, *Ind. Eng. Chem. Res.* **46**, 43 (2007).
8. T. Doll and F. Frimmel, *Water Res.* **38**, 955 (2004).
9. C. S. Turchi and D. F. Ollis, *J. Catal.* **122**, 178 (1990).
10. C. B. Almquist and P. Biswas, *Chem. Eng. Sci.* **56**, 3421 (2001).
11. E. Pelizzetti and C. Minero, *Electrochim. Acta* **1993**, 38 (1993).
12. M. F. J. Dijkstra, H. J. Panneman and J. G. M. Winkelman, *Chem. Eng. Sci.* **57**, 4895 (2002).
13. R. Terzian, N. Serpone, C. Minero, E. Pelizzetti and H. Hidaka, *J. Photochem. Photobiol. A* **55**, 243 (1990).
14. A. Mills, S. Morris and R. Davies, *J. Photochem. Photobiol. A* **70**, 183 (1993).
15. J. Theurich, M. Lindner and D. W. Bahnemann, *Langmuir* **12**, 6368 (1996).
16. O. M. Alfano, M. I. Cabrera and A. E. Cassano, *J. Catal.* **172**, 370 (1997).
17. G. Spadoni, E. Bandini and F. Santarelli, *Chem. Eng. Sci.* **33**, 517 (1977).
18. T. Yokota, S. Cesur, H. Suzuki, H. Baba and Y. Takahata, *J. Chem. Eng. Jpn.* **32**, 314 (1999).
19. J. Moreira, B. Serrano, A. Ortíz and H. de Lasa, *Ind. Eng. Chem. Res.* **49**, 10524 (2010).
20. M. Zekri and C. Colbeau-Justin, *Chem. Eng. J.* **225**, 547 (2013).
21. S. L. Murov, I. Carmichael and G. L. Hug, *Handbook of Photochemistry* (Marcel Dekker, New York, 1993).

Chapter 5

Efficiency Versus Productivity in Photoreactors: A Case Study

M. Enis Leblebici, Bart Van den Bogaert, Georgios D. Stefanidis
and Tom Van Gerven*

Process Engineering for Sustainable Systems (ProcESS)
Department of Chemical Engineering, KU Leuven, Belgium
*tom.vangerven@kuleuven.be

5.1 Introduction

Flow chemistry has been going hand-in-hand with photochemistry for some time already.[1] The first and main reason for this collaboration is the fact that the photochemical reactor productivity increases exponentially with the increasing rate of absorbed radiative energy, which drives the reaction forward.

In most photochemical and photocatalytic degradation reactions, the stoichiometry of photons to products is one.[2,3] In other words, one photon is theoretically necessary to excite one photosensitive molecule, which will then undergo either an oxidative or a reductive quenching towards a new synthesis or a degradation route as shown in Figure 5.1.[3]

However, the photosensitive molecule has a very short lifetime at its excited state and it recombines to the ground state in less than nanosecond timescales.[2] This high reactivity of photocatalysts has the potential of depleting the substrate reagents in reactor zones where the photon flux is high, forcing the reaction mechanism towards undesired

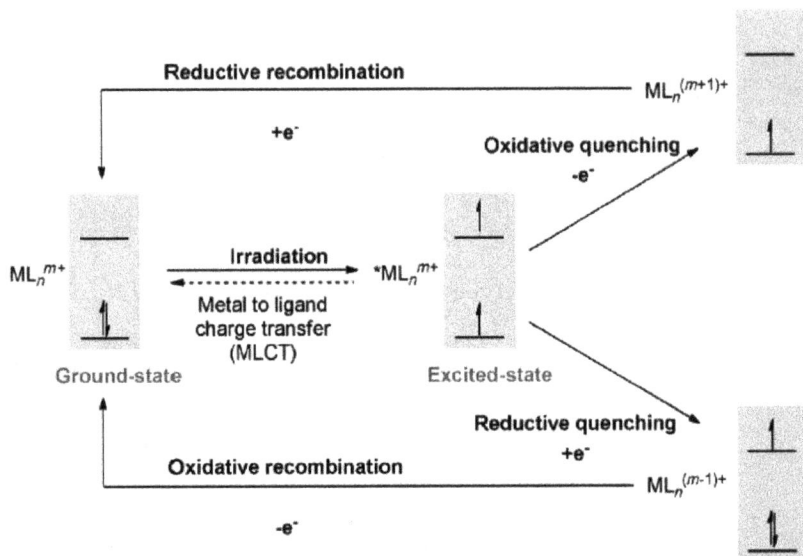

Fig. 5.1. Reaction mechanisms in a photocatalytic reactor. (Reprinted with permission from Ref. 3. Copyright © 2014 Wiley-VCH Verlag GmbH & Co. KGaA, Weinheim.)

recombination as previously shown by Leblebici *et al.*[4] In their phenol degradation reactor modeling paper, CFD tools have been used to couple irradiance intensity to $^{\bullet}$OH radical generation rate to prove the existence of depleted zones in the titania coating of a parallel plate photocatalytic reactor, as shown in Figure 5.2.[2]

Figure 5.2(b) shows that the light can penetrate throughout a catalyst coating. However, due to the diffusion limitations, only a third of the catalyst can be utilized in this particular phenol degradation reactor. The existence of these depleted zones decreases the reaction rate and intensifies the loss of photonic efficiency due to the mass transfer limitations. In a standard catalytic reactor, mass transfer limitations at the porous catalyst surface are common. This phenomenon is always evidenced by a steep substrate concentration gradient within the porous medium as shown in Figure 5.2(a). This is unwanted since it means where there is valuable catalyst, there is less

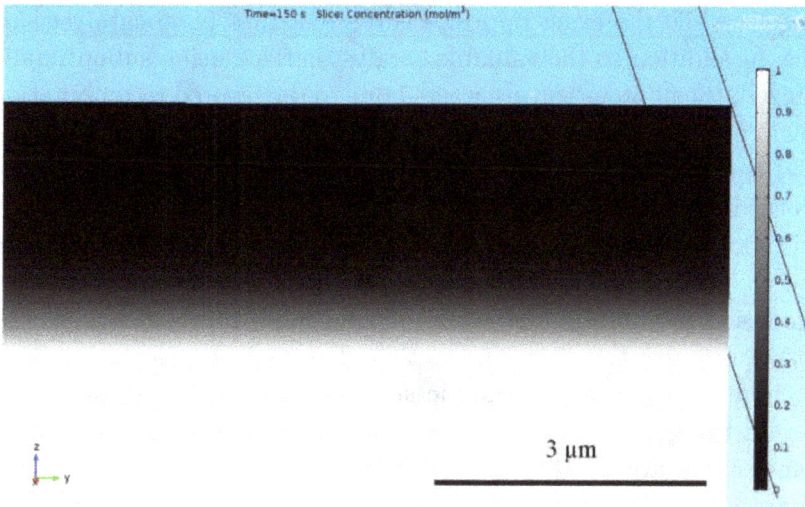

(a)

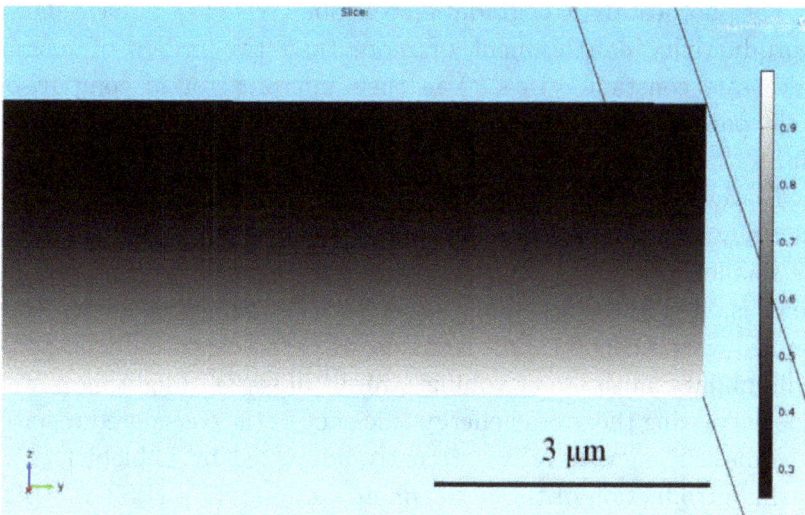

(b)

Fig. 5.2. Calculated concentration (a) and irradiance (b) gradients in a catalyst coating. Both concentration and irradiance values are relative to their values at catalysts coating — reaction medium interface. (Reprinted with permission from Ref. 2. Copyright © 2015 Elsevier.)

reagent while the exact opposite is desired. In a photocatalyst surface, in addition to the valuable catalyst surface being suboptimally utilized, the photons too are wasted due to the favored recombination reaction in depleted zones.

Reactor designers have been trying to overcome these limitations via the instruments of process intensification for some time already.[5,6] One tool to attack this problem was previously shown as the micro flow chemistry.[7-9] The microreactor technology decreased the reactor dimension to sub-mm and enabled rapid mass and heat transfer by both shortening the diffusion length and also introducing new mixing mechanisms by means of chaotic advection e.g. the Taylor flow.

On this foundation of flow chemistry, it was shown that photon transfer limitations were also addressed as well as the mass transfer issues previously stated. Significant developments on the scale-up of microreactor cascades are also being made.[1]

For photocatalytic degradation reactions, Visan *et al.* have shown a productivity enhancement of more than two orders of magnitude (rate constant ~ 15 s^{-1}) in their microreactor in comparison to the parallel plate reactors with 3 mm flow thickness (rate constant ~ 0.1 s^{-1}).[2,7]

However, high productivity does not always mean high energy efficiency. Regardless of the high space–time yields (STYs) achieved, the smaller volume of microreactors usually means lower throughput, which is often in the mL min^{-1} scale. This low production rate, when coupled with the light source technologies, which are usually designed to illuminate much larger volumes, result in waste of lighting power, thus decreasing the overall energy efficiency of the considered reactor.

This efficiency issue was recently addressed by Leblebici *et al.* by the introduction of the term, photocatalytic STY (PSTY).[9] The following section will elaborate further on this study.

5.1.1 *Productivity versus energy efficiency*

In order to assess the productivity and energy efficiency of different photoreactor designs, a solid benchmark is required. Two benchmarks have been widely used in the relevant literature. The first one is the apparent first-order reaction rate constant, k. It is usually expressed in terms of min^{-1} and gives a direct view on the conversion rate, which is quite useful when comparing different reactors with similar volumes. However, it gives no information on throughput since it is volume-dependent. For example, a parallel-plate reactor working as a loop will give a much lower k when connected to a 10-L vessel than a 50-mL vessel although it will perform just the same since the active area does not vary. Furthermore, in some photochemical systems, the photon field is the limiting "reagent", which results in a zero-order kinetics as evidenced by Van den Bogaert et al.[10] This study will be elaborated in later sections. This benchmark is also light intensity- and catalyst load-dependent.

The second popular benchmark is the photonic efficiency, also known as the quantum yield, which is expressed as

$$\varepsilon = z\frac{R}{\Phi} \times 100 \tag{1}$$

where ε is the dimensionless photonic efficiency, R is the reaction rate $(mol\,L^{-1}\,s^{-1})$, z is the amount of electrons transferred per molecule to be degraded or converted and Φ is the photon flux $(mol\,L^{-1}\,s^{-1})$.

ε expresses the light utilization efficiency of the reactor. However, neither ε nor k provide information on the electrical consumption or productivity of the reactor. As an example, an annular reactor and a parallel-plate reactor may have the same photonic efficiency[2] and/or the same k; however, the two reactors may need lamps of very different powers and may work at different throughput levels.

The simple benchmark introduced by Leblebici et al.[9] is the ratio of STY to the standardized electricity expenditure of lamp. STY is the measure of productivity in reactor design. It can be calculated by taking the inverse of passage time in flow reactors. The STY for water treatment (degradation) reactors can be standardized to the volume (m^3) of the wastewater processed from e.g. $100\,mmol\,L^{-1}$ of pollutant to $0.1\,mmol\,L^{-1}$ in a day by the reactor when it is scaled up to $1\,m^3$. In flow systems with first-order kinetics, STY can be calculated easily from the apparent reaction rate constant. For a continuous slurry reactor and for continuous immobilized catalyst reactor operated in a loop, the following stirred tank reactor (CSTR) model equation can be used to predict the outlet concentration:

$$C_A = \frac{C_{A0}}{1 + k\tau} \tag{2}$$

where C_A is the outlet concentration in $\mathrm{mmol\,L^{-1}}$, C_{A0} is the inlet concentration in $\mathrm{mmol\,L^{-1}}$ and τ is the passage time in days. From the CSTR model, τ can be calculated. Replacing C_A with 0.1 $\mathrm{mmol\,L^{-1}}$ and C_{A0} with 100 $\mathrm{mmol\,L^{-1}}$ yields

$$STY_{cstr} = \frac{V_R}{\tau} = \frac{1\,\mathrm{m}^3}{\tau} = \frac{k}{999} \tag{3}$$

where V_R is the reactor volume $(1\,\mathrm{m}^3)$ and k has units of $\mathrm{day^{-1}}$. The inlet and outlet concentrations were selected to standardize the degradation to three orders of magnitude of concentration decrease. This is a common conversion standard, especially in pharmaceutical and insecticide pollutant elimination.[11] For plug-flow photocatalytic reactors, such as the microreactor, the STY can be expressed as

$$STY_{pfr} = \frac{1\,\mathrm{m}^3}{\tau} = -\frac{k}{\ln\frac{C_A}{C_{A0}}} = \frac{k}{6.908} \tag{4}$$

For the case of this work, benchmarking a batch reactor will be necessary. Furthermore, the reactor will not work for wastewater treatment. Instead, there will be a product formation. Hence, for this case, STY can be calculated using;

$$STY_{batch} = M_w \left(\frac{dC}{dt}\right) \tag{5}$$

where M_w is the molecular weight $(\mathrm{g\,mol^{-1}})$. It can be seen that the STY $(\mathrm{g\,L^{-1}\,day^{-1}})$ is a different form of benchmark than ε and k. It can, however, be derived from k by the aid of V_R. STY only gives a measure of productivity. Therefore, it needs to be enhanced to include the lighting energy efficiency to the benchmark. To do this, the calculated STY was divided by the lamp power (LP). The

LP was scaled to the value, which would illuminate one unit volume (L in this case) of the reactor by the following relation:

$$LP = P \times \frac{1\,L}{V} \qquad (6)$$

where LP is the standardized lamp power (kW or W per L), P is the lamp power of the experimental setup and V is the volume of the reaction medium in the experimental setup (L).

Hence, the overall PSTY is defined as

$$PSTY = \frac{STY}{LP} \qquad (7)$$

The productivity of a photoreactor therefore can be assessed via the STY (mass or volume of product per time per volume of reactor) and the energy efficiency can be assessed via the PSTY (mass or volume of product per time per unit energy input per volume of reactor).

There are four main aspects to consider, which may affect these parameters in different reactor geometries and reaction schemes. These aspects are:

(1) the local volumetric rate of radiative energy absorption (LVREA);
(2) the outward photon flux;
(3) the mechanism of the photochemical reaction;
(4) the presence and physics of solids.

The STY and the PSTY are dependent on these aspects.

The LVREA is the total amount of energy absorbed via radiation at any given point in a reactor per unit time ($W\,m^{-3}$). It can be given by

$$LVREA_\lambda(s, t) = G_\lambda(s, t)k_{a,\lambda} \qquad (8)$$

where G ($W\,m^{-2}$) is the scalar incident irradiation for wavelength (λ) at any position (s) at any time (t) and k_a is the absorption coefficient in m^{-1} for wavelength (λ). Calculation of the $G_\lambda(s, t)$ requires the solution of the radiative transfer equation (RTE), which is further elaborated on in the modeling section of this work. The LVREA has a direct effect on productivity. This term can also be converted to the volumetric rate of radiative energy absorption (VREA), which is the integrated form of LVREA throughout the reactor volume and it bears the term W. The VREA can also be converted to the rate of photon absorption via:

$$\Phi = \frac{\text{VREA}}{Nh\left(\frac{c}{\lambda}\right)} \qquad (9)$$

where Φ is the absolute rate of photon absorption in $mol\,s^{-1}$, N is the Avogadro number in mol^{-1}, h is the Planck constant in $J\,s$ and c is the speed of light in $m\,s^{-1}$. This way, the photonic efficiency (equation (1)) can be calculated.

Intuitively, the microreactor represented by Figure 5.3(a) has the advantage of keeping all of the reaction medium very close to the light source, resulting in significantly higher LVREA, which results in a much higher STY compared to a reactor with significant light-path length (Figure 5.3(b)).

Outward photon flux, in contrast to LVREA, has no effect on the productivity (STY). These wasted photons, however, will decrease the energy efficiency (PSTY). Although the microreactor performs better in terms of STY, its PSTY is significantly lower than competing geometries. This is due to the LP parameter (equation (6)), which would be higher for a smaller reactor volume keeping the LP constant. This was observed by Leblebici *et al.* in their work regarding PSTY.[9] In the said work, the microreactor introduced by Visan *et al.*[7] had two to five orders of magnitude higher productivity than the rest of the benchmarked reactors. However, due to the extremely large LP value (8×10^7 kW), its efficiency (PSTY) was compromised. In photocatalytic wastewater treatment, all the industrially applied designs have dramatically low LP values ranging from 0.22 to $4\,kW\,m^{-3}$.

The *mechanism* of the involved photochemical reaction is another important parameter governing the reactor geometry. Particularly if there are reverse reactions, either catalyzed by or independent of light, the mechanism becomes of higher importance. Higher the

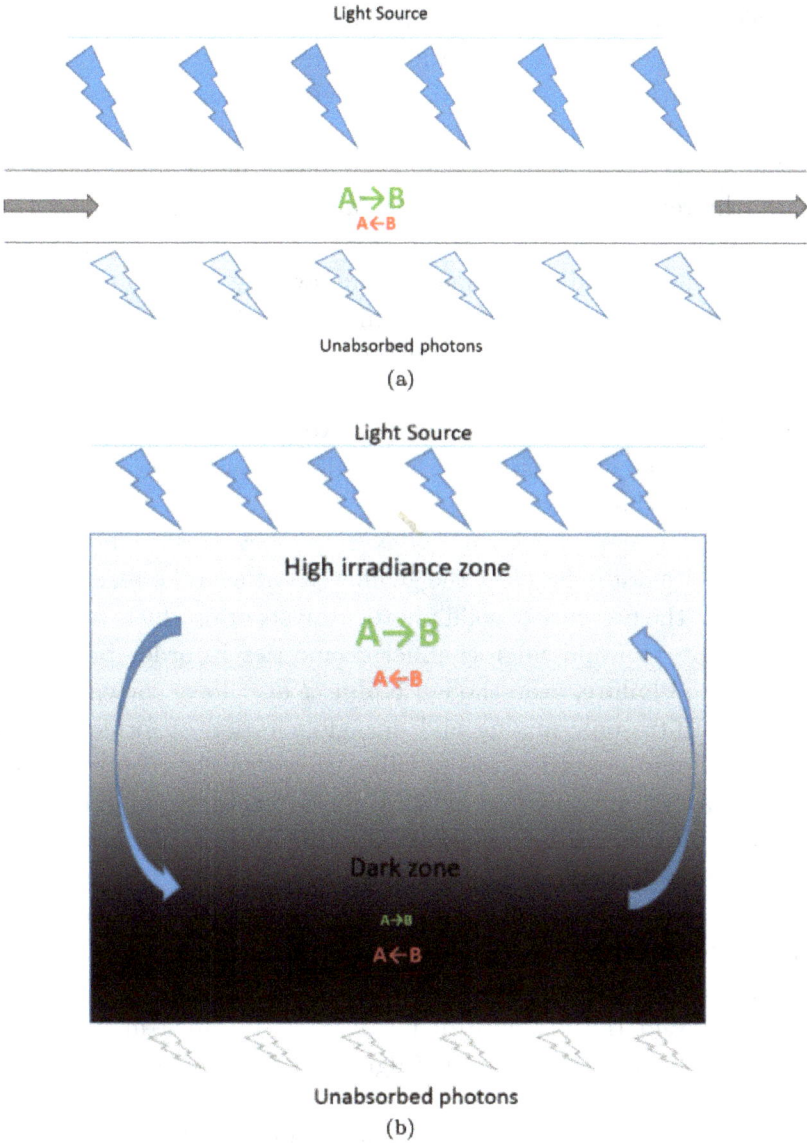

Fig. 5.3. Different mechanisms in (a) a microreactor and (b) a batch or flow reactor with significant light-path length. (Reprinted with permission from Ref. 39. Copyright © 2016 Elsevier B.V.)

backward reaction rate is, smaller will be the light-path length for optimum efficiency. The reason for this design rule can be observed in Figure 5.3(b) where there are both high-irradiance zones and low-irradiance (dark) zones due to the light gradient in high-absorbing media. While the forward reaction is dominant in the high-irradiance zones, the reverse reaction would be favored in the dark zones. Due to the vortices created by mixing in the reactor, the reagents will undergo a periodic illumination scheme favoring the two counter — acting mechanisms consecutively. This would decrease and in some cases even cease the productivity (and hence the efficiency) of the reactor. Therefore, if there are prevalent reverse reaction mechanisms, the poorly illuminated zones have to be minimized or prevented. This can even justify the "wasted" outward photon flux using a milli/micro channel reactor.

Formation and physics of solids as photocatalysts and reaction products does not only affect the geometry but also the reactor type. Intuitively, the presence of solid photocatalysts or products may favor the decision to avoid microchannel geometries in order to prevent clogging and failure, even though Kuhn *et al.*[12] have shown that by the aid of ultrasonic mixing and deagglomeration of formed solids, the solid–liquid operation was possible. However, the required power for such a system would total $30\,\mathrm{W}$ for a reactor volume of $1\,\mathrm{mL}$, corresponding to an LP value of $30 \times 10^3\ \mathrm{W\,m^{-3}}$. This would be justified only if the small geometry is also justified by the design principles stated above. Apart from the clogging risk, the solids also cause light interference, for instance due to scattering and absorption. There are several analytical and numerical models to calculate the LVREA in the presence of particles,[13–15] which may need to be applied to calculate the new light distribution after the solids formation.

As can be seen from the five mechanisms introduced above, photoreactor design is an optimization problem balancing between efficiency and productivity. Micro/milli channels have a high productivity. However, they can exhibit low energy efficiency due to a large outward photon flux. On the other hand, reactors with significantly large light-path lengths may have a very low rate of productivity due to the presence of solids and/or back reactions.

The aim of this work is to illustrate this optimization problem with a case-study on photochemical recycling of rare-earth elements (REEs) and also to propose a methodology to obtain the optimum reactor geometries for different cases.

5.2 Photochemical Recycling of Rare Earths from Lamp Phosphors

5.2.1 *Rare-earth lamp phosphors*

The REEs are a group of seventeen chemical elements consisting of the fifteen lanthanides (from lanthanum (atomic number 57) up to lutetium (atomic number 71)), supplemented with scandium and yttrium. Many of the REEs are used in high-tech and green applications, such as permanent magnets in wind turbines, batteries in hybrid and electrical vehicles and lamp phosphors in compact fluorescent lamps (CFLs).[16] Due to the quasi-monopoly of China in mining and downstream processing of rare earths, these elements are scarce and valuable on the global market.[17] Because of their high economic importance, especially in green technology, and their high supply risk, the REEs are considered critical raw materials.[18] Therefore, rare-earth recycling from end-of-life products has become increasingly important to maintain a reliable supply of REEs, especially for countries or regions lacking rare-earth resources.[19] In this case study, emphasis is put on europium and yttrium, two critical REEs. The main application of these two elements is found in lamp phosphors used in CFLs, more specifically as the red phosphor YOX (Y_2O_3:Eu^{3+} or yttria doped with europium(III)). In CFLs, a triphosphor blend consisting of a green, blue and red phosphor creates the white light output as perceived by the human eye. All three phosphors contain REEs, a typical phosphor composition is listed in Table 5.1. The phosphors consist of around 15% of REEs and make up about 3% of the total weight of a CFL.[20]

The red phosphor YOX is the main component in the triphosphor blend and purely consists of critical REEs, namely yttrium

Table 5.1.   Typical composition of a triphosphor blend in CFLs.

Color	Phosphor name	Phosphor composition	wt% in the blend (%)
Red	YOX	$Y_2O_3{:}Eu^{3+}$	50
Green	LAP	$LaPO_4{:}(Ce^{3+},Tb^{3+})$	35
Blue	BAM	$BaMgAl_{10}O_{17}{:}Eu^{2+}$	15

and europium. Therefore, this could serve as an interesting waste stream to recycle europium and yttrium from. The main challenge is to develop a selective and efficient separation technique to separate europium and yttrium into pure fractions. Due to the very similar physical and chemical properties of REEs, conventional separation methods such as solvent extraction and ion exchange have very low separation factors ($\beta = 2.5$ for neighboring REEs).[21] This demands many extraction stages to reach the purities desired in consumer products (>99.99%).[22]

5.2.2   *Redox behavior of REEs*

For the REEs, the most common and stable oxidation state in aqueous solutions is the trivalent ($+3$) state. However, some elements, such as europium, can adopt a divalent ($+2$) oxidation state, while others, such as cerium, can form stable tetravalent ($+4$) species.[23] Separations based on oxidation state therefore show great potential since they are much more selective toward one single element, which allows to efficiently remove this particular element from a rare-earth mixture. This results in a highly selective one-step separation process, which is particularly interesting for binary mixtures consisting of one element prone to redox reactions and one element immune for redox reactions. An example of such a binary mixture is europium/yttrium, where europium can be selectively reduced to its divalent state while yttrium remains in its trivalent form.[10] Subsequently, the divalent europium is removed from the solution, e.g. by precipitation with sulfate. Effective separation of europium from yttrium with much higher selectivity and efficiency than conventional separation methods can be performed this way.

In order to exploit the characteristic redox properties of europium for the separation of the said binary mixture, europium reduction has to take place. This is typically performed chemically,[24] electrochemically[25] or photochemically.[10,26,27] Chemical reduction consumes zinc powder or zinc amalgam, which pollutes the end product, thus requiring extra purification steps. Electrochemical techniques using graphite or titanium electrodes, on the other hand, have low yields due to hydrogen evolution by a side reaction. Photochemical reduction results in the highest selectivity and consumes less harmful and toxic chemicals. In this process, photons, which are emitted by a UV light source, assist in an electron transfer from the environment toward europium(III), which in turn causes the reduction of europium(III) to europium(II). Subsequently, the divalent europium is removed from the mixture by precipitation as $EuSO_4$. Different light sources have been used to obtain the photochemical reduction, such as excimer lasers,[28] high-pressure mercury lamps (HPMLs)[29] and low-pressure mercury lamps (LPMLs).[10,26] In this study, a U-shaped 160-W LPML was used. The main photochemical processes influencing the reduction of europium are explained below.

5.2.3 *Photoreduction of Eu^{3+} in water*

When a rare-earth mixture containing europium(III) is illuminated with a UV light source, photons are absorbed by the solution. These photons cause a ligand-to-metal charge transfer (LMCT), during which an electron from the solution is transferred to europium(III), resulting in a reduction to europium(II). In water, this LMCT takes place at an energy corresponding to light of 188 nm.[27]

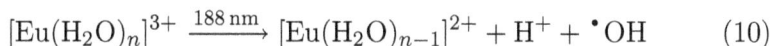

$$[Eu(H_2O)_n]^{3+} \xrightarrow{188\,nm} [Eu(H_2O)_{n-1}]^{2+} + H^+ + {}^{\bullet}OH \qquad (10)$$

During this reduction, a proton and a hydroxyl radical are formed. The yield of equation (10) can be maximized in three ways: (1) removal of the main product, i.e. the reduced species Eu^{2+}, (2) destruction of the ${}^{\bullet}OH$ radicals and (3) control of the proton concentration or, in other words, the pH. In the following sections, these three measures are discussed more in detail.

Table 5.2. Solubility of trivalent and divalent rare-earth sulfates in water.

Compound	Solubility (g/100 g H_2O)
Yttrium(III) sulfate	7.47
Europium(III) sulfate	2.10
Europium(II) sulfate	<0.001

Removal of Eu^{2+}

Firstly, the yield of this reaction can be maximized by removing the divalent species from the solution. An efficient way to accomplish this is by adding sulfates (equation (11)), since the solubility of trivalent rare-earth sulfates is much higher than the sparingly soluble europium(II) sulfate (see Table 5.2)[30]:

$$Eu^{2+} + SO_4^{2-} \rightarrow EuSO_{4,s} \quad K_s = 1.5 \times 10^{-9} \tag{11}$$

Not only does the addition of sulfates cause europium(II) to precipitate, it also creates an extra charge transfer band from sulfate-to-europium(III), which results in an extra reduction of europium(III) to europium(II) (equations (12) and (13))[26]:

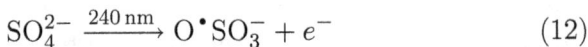

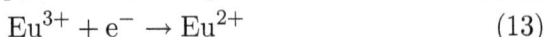

$$SO_4^{2-} \xrightarrow{240\,nm} O^{\bullet}SO_3^{-} + e^{-} \tag{12}$$

$$Eu^{3+} + e^{-} \rightarrow Eu^{2+} \tag{13}$$

A sulfate excess of five–seven times with regard to the initial europium(III) concentration yields maximum europium removal. A further increase of the amount of sulfates has no more beneficial impact on the recovery rate. The sulfates can be added as Na_2SO_4, $(NH_4)_2SO_4$, K_2SO_4 or any other sulfate salt without changing the removal efficiency. The radicals formed in equation (12) readily react with water to form hydroxyl radicals and bisulfate ions (equation (14))[10]

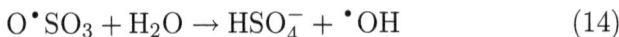

$$O^{\bullet}SO_3 + H_2O \rightarrow HSO_4^{-} + {}^{\bullet}OH \tag{14}$$

Radical scavenging

Not only are hydroxyl radicals formed in equation (1), they are also a by-product from the main reduction in equation (1), and should be removed from the reaction medium, since they are very reactive toward europium(II), causing oxidation back to europium(III) (equation (15))[31]:

$$Eu^{2+} + {}^{\bullet}OH \rightarrow Eu^{3+} + OH^- \tag{15}$$

This unwanted oxidation can be avoided by the addition of radical scavengers, typically organic compounds such as alcohols, esters or carboxylic acids. They react with the hydroxyl radicals and form more stable organic radicals, neutralizing the oxidative power of the hydroxyl radicals. The most commonly used scavenger is isopropanol, and its scavenger reaction is shown in equation (16)[32]:

$$(CH_3)_2CHOH + {}^{\bullet}OH \rightarrow (CH_3)_2C^{\bullet}OH + H_2O \tag{16}$$

As a bonus, the organic radicals formed in the reaction above have the ability to reduce europium(III) to europium(II), which can be given as[32]:

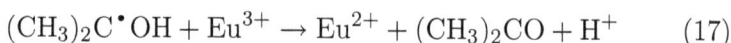

$$(CH_3)_2C^{\bullet}OH + Eu^{3+} \rightarrow Eu^{2+} + (CH_3)_2CO + H^+ \tag{17}$$

Since hydroxyl radicals are very reactive, large amounts of scavengers have to be added to the solution, up to an isopropanol/europium molar ratio of 250/1.[10]

pH optimization

Next to hydroxyl radicals, a second by-product is formed in equation (10), namely protons. In order to maximize the yield of europium(II), the proton concentration should be kept low.[10] This will draw the reaction equilibrium towards the side of the divalent species. Moreover, high proton concentrations induce unwanted oxidation of europium(III) to europium(II), either directly[33] (equation (18)) or photochemically[34] (equation (19)).

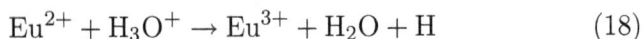

$$Eu^{2+} + H_3O^+ \rightarrow Eu^{3+} + H_2O + H \tag{18}$$

$$2\ Eu^{2+} + 2\ H^+ \xrightarrow{366\,nm} 2\ Eu^{3+} + H_2 \tag{19}$$

The share of equation (19) can also be minimized by avoiding the presence of light with the wavelength of 366 nm, by using either band-pass filters or monochromatic light sources. The LPML used in this study, however, partially emits light of 366 nm, which could cause some photochemical back-oxidation.

Equations (10), (18) and (19) suggest to work at low proton concentration, i.e. at high pH. Furthermore, Eu^{2+} is thermodynamically more stable at less acidic conditions, as determined by the Nernst equation, which describes the pH dependence of water stability (equation (20))

$$E_{H^+/H_2} = E^0_{H^+/H_2} + \frac{RT}{nF} \ln\left(\frac{[H^+]^2}{p_{H_2}}\right) \tag{20}$$

At standard conditions ($T = 298$ K, $p = 1$ atm), this can be written as

$$E = -0.059\ pH \tag{21}$$

The standard reduction potential of Eu^{3+}/Eu^{2+} in water is -0.34 V, which results in a thermodynamically stable species of Eu^{2+} above pH 5.8.[35]

However, there is a limit on increasing the pH value since in alkaline media (pH < 7), all trivalent REEs undergo hydrolysis (equation (22)) to form insoluble hydroxides:

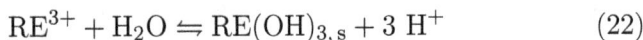

$$RE^{3+} + H_2O \rightleftharpoons RE(OH)_{3,\,s} + 3\ H^+ \tag{22}$$

This results in unselective precipitation of all REEs and no separation into pure fractions can be obtained. Therefore, a moderately acidic medium is preferred for the photochemical reduction of europium(III), with the optimum conditions at a pH of 3.9.[10]

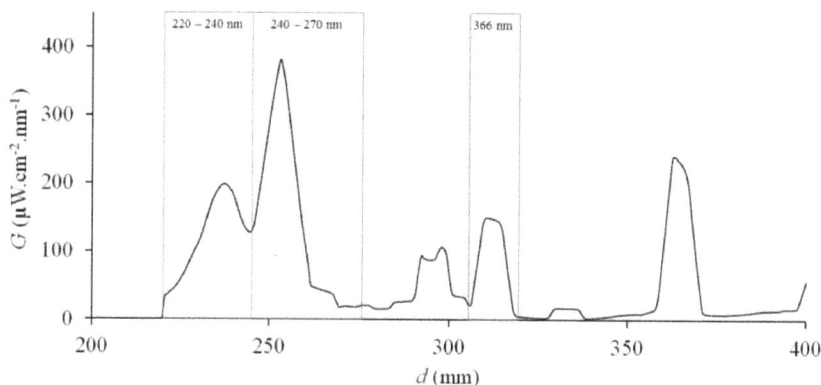

Fig. 5.4. Output spectrum of a U-shaped 160-W LPML at a distance of 25 mm. (Reprinted with permission from Ref. 39. Copyright © 2016 Elsevier B.V.)

5.2.4 *Summary*

It is clear that the chemical parameters have to be controlled carefully in order to maximize the yield of europium(II). In this case study, the optimal parameters, as determined by Van den Bogaert *et al.*[10] were used. These parameters consisted of 10 mM $EuCl_3 \cdot 6H_2O$, a five-time excess (50 mM) of $(NH_4)_2SO_4$ and a 250-time excess of isopropanol as a scavenger in an aqueous medium of pH 3.9. An LPML source is used, since its output spectrum coincides with the requirements for europium(III) reduction, i.e. around 188 nm (see equation (10)) and around 240 nm (see equation (12)). The unwanted output around 366 nm (see equation (19)) is not completely absent, but much less intense than the wavelengths corresponding to the forward reactions (see Figure 5.4).

5.3 Materials and Methods

Two types of experiments were conducted for this study. Firstly, deep batch experiments (76 mm and 40 mm depth) were carried out to obtain kinetics with standard batch reactors, as well as shallow batch experiments (5–30 mm) to emulate millichannel geometries and measure the productivity at shorter light paths. Secondly, various irradiance measurements were conducted to validate the radiative transfer model, which is utilized to calculate the LVREA for shallow batch reactors.

The *reaction medium* for all experiments was the same. Europium was added as its chloride hexahydrate salt ($EuCl_3 \cdot 6H_2O$) and had a purity of 99.9% (Acros Organics). Ammonium sulfate (99.6%, Acros Organics) was added in its solid form. Isopropanol (99.5%, VWR) was used as a radical scavenger. A 1-M HCl stock solution (Fluka Chemicals) was used to obtain the desired pH of 3.9 by diluting it with ultrapure water (Milli-Q).

Deep batch experiments were conducted by the method published by Van den Bogaert et al.,[36] which uses two types of jacketed cylindrical borosilicate reactor vessels. The first one has a diameter of 4 cm with a total volume of 100 mL (height = 76 mm). The second type of reactor had a diameter of 5 mm and is filled to the height of 40 mm. The reactors were covered with rectangular $5 \times 5 \, cm^2$ quartz glasses, which are transparent for wavelengths longer than 200 nm. A cooling bath (Julabo F12-ED) applied cooling to 20°C. The solution was magnetically stirred during the reaction. The 160-W LPML was placed horizontally above the reactor vessels to provide illumination from the top. The distance between the solution surface and the arc of the lamp was varied for different experiments. The setup is depicted in Figure 5.5.

Shallow batch experiments were conducted using a shorter jacketed cylindrical reactor, which was made completely from quartz (diameter = 39 mm). The 160-W LPML was placed horizontally below the reactor. The distance between lamp and the reactor bottom was kept constant at 2 mm. The reactor was filled with liquid with depth varying from 5 mm (6.1 mL) to 30 mm (36.6 mL). It was stirred with an agitator at 150 rpm. The reactor was cooled to 5°C to avoid evaporation. The setup is depicted in Figure 5.6.

Both the setups were constructed in a sealed dark box to protect the surroundings against the hazardous UV radiation. Special UV protective goggles (LOT-Oriel) were worn during the experiment. Solutions were illuminated for several hours and samples were taken at various time intervals in order to determine the rare-earth concentration in solution at different illumination times. An aliquot

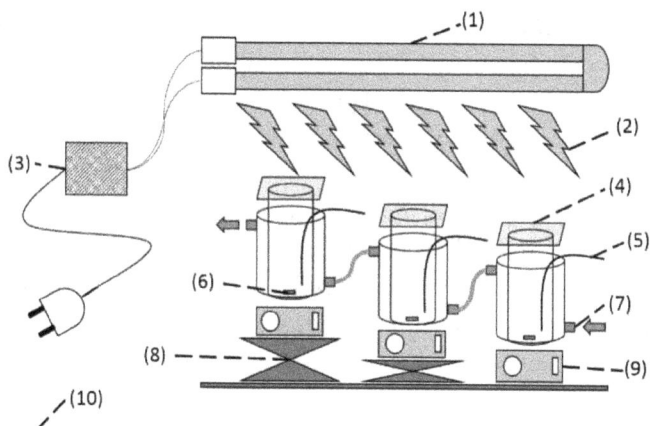

Fig. 5.5.   Setup for deep batch experiments. (1) 160-W U-shaped LPML, (2) UV irradiation, (3) electronic ballast, (4) quartz covering plate, (5) sampling tube, (6) magnetic stirring bar, (7) cooling jacket (arrows indicate water flow), (8) height adjuster to vary lamp-to-solution surface distance, (9) magnetic stirring plate and (10) protective dark box. (Reprinted with permission from Ref. 36. Copyright © 2016 RSC publishing.)

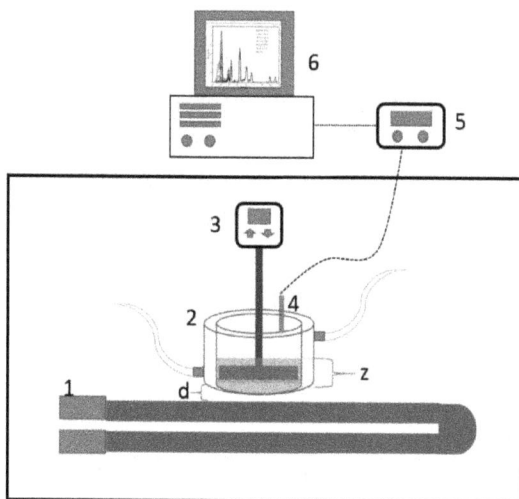

Fig. 5.6.   Setup for deep batch experiments consisting of a 160-W LPML (1), quartz reactor vessel (2), stirrer (3), spectrometer probe and optical fiber (4), spectrometer (5) and the data acquisition computer (6). The distance from lamp (d) was constant at 2 mm and the liquid depth (z) was varied between 5 and 30 mm. (Reprinted with permission from Ref. 39. Copyright © 2016 Elsevier B.V.)

of the sample solution (0.5 mL for deep batch and 0.1 mL for shallow batch) was centrifuged at 4000 rpm for 10 min in an Eppendorf tube of 1.6 mL to separate any $EuSO_4$ precipitation from the supernatant. An aliquot of the supernatant was taken for analysis of the metal content.

Analysis of the metal concentration of the liquid phase was performed by total-reflection X-ray fluorescence (TXRF) spectrometry, using a Bruker S2 Picofox TXRF spectrometer. An aliquot of the sample (100 μL) was mixed with 100 μL of a 1000-mg/L gallium internal standard solution and diluted with 800 μL of ultrapure water (Milli-Q). A droplet of 7 μL was put on a quartz sample carrier, which was pre-coated with a silicon solution in isopropanol (SERVA) to make the carrier hydrophobic in order to avoid spreading of the sample droplet on the carrier. The quartz glasses were dried in an oven at 60°C for 30 min and analyzed with the TXRF spectrometer.

5.4 Model

In the introduction section, the LVREA was given as one of the key factors determining the photoreactor performance. It was also stated that modeling the light distribution within the reactor was necessary, in order to calculate the LVREA as well as the photon flux and to evaluate their interactions with the STY and PSTY terms.

For the model, COMSOL Multiphysics 5.2 radiation heat transfer node was utilized along with the mathematics node. First, the experimental setup was recreated in the software as shown in Figure 5.7(a). Afterwards, the governing equation to solve was determined.

As explained in the introduction section, it is required to solve the RTE to calculate the spectral irradiance and finally the LVREA. The RTE can be given as[37]:

$$\Omega \cdot \nabla I_\lambda(\Omega) = j_\lambda(\Omega) - \kappa_a I_\lambda(\Omega) + \frac{\kappa_s}{4\pi} \int_{4\pi} I_\lambda(\Omega') f(\Omega, \Omega') d\Omega'$$

$$(23)$$

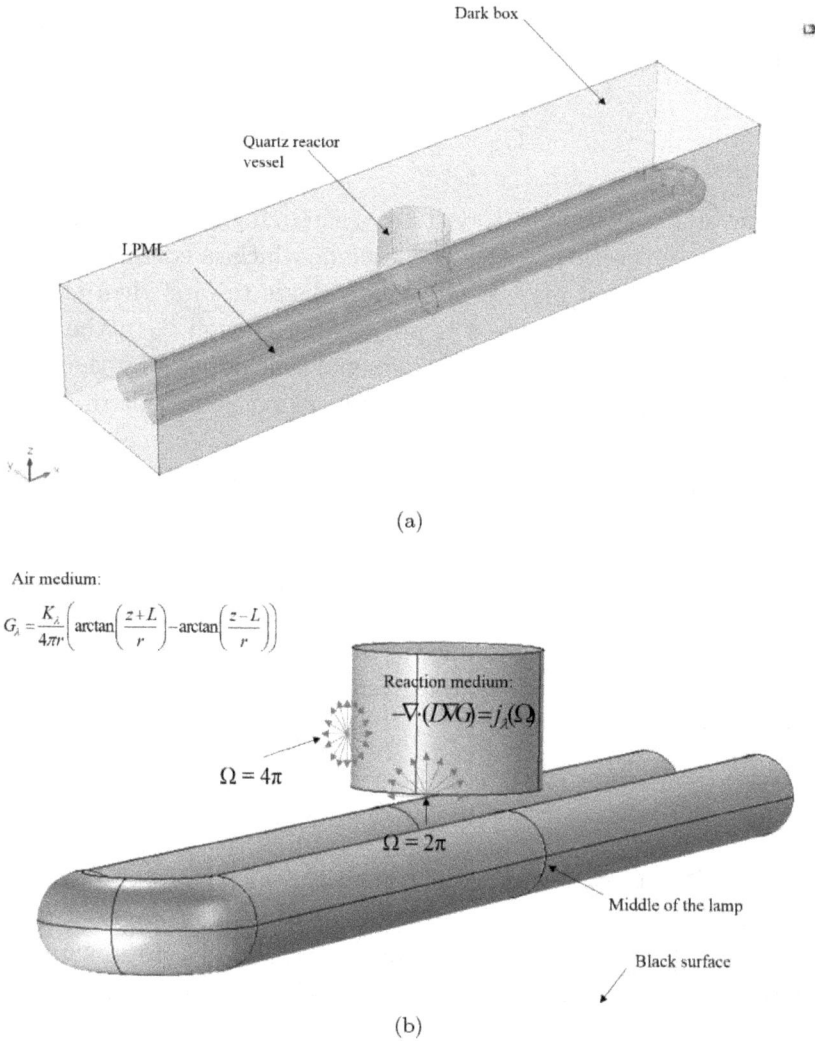

Dark box

Quartz reactor
vessel

LPML

(a)

Air medium:

$$G_\lambda = \frac{K_\lambda}{4\pi r}\left(\arctan\left(\frac{z+L}{r}\right) - \arctan\left(\frac{z-L}{r}\right)\right)$$

Reaction medium:

$$-\nabla \cdot (D\nabla G) = j_\lambda(\Omega)$$

$\Omega = 4\pi$

$\Omega = 2\pi$

Middle of the lamp

Black surface

(b)

Fig. 5.7. Simulation domains with the lamp and reactor (a), governing equations with the boundary conditions (b). (Reprinted with permission from Ref. 39. Copyright © 2016 Elsevier B.V.)

where I_λ is the radiance ($\mathrm{W\,m^{-2}\,sr}$) at the wavelength λ, j_λ is the emission coefficient vector ($\mathrm{W\,m^{-2}\,sr}$), κ_a and κ_s are the absorption and scattering coefficients ($\mathrm{m^{-1}}$), respectively, and the Ω denotes the solid angle (sr). By solving the RTE, it is possible to calculate the scalar $G(s,t)$ quantity via integrating all the radiance vectors around a spherical solid angle:

$$G_\lambda = \int_{4\pi} I_\lambda(\Omega)d\Omega \qquad (24)$$

However, RTE is usually solved by discretizing the solid angle at every discrete position in the 3D domain, which is computationally very demanding and may result in unrealistic results when the discretization of the solid angle is not refined enough.[38] In the trials using the discrete ordinates, the refinement achievable by the existing computers in expense of this study was not sufficient to yield a valid model.

For this reason, the P1 approximation of the RTE was utilized. The simplified equation can be given by

$$-\nabla \cdot (D\nabla G) = j_\lambda(\Omega) \qquad (25)$$

where G is the spectral irradiance as in equation (8) and the D is the diffusion coefficient, which can be given assuming isotropic scattering and absorption as follows:

$$D = \frac{1}{3\kappa_a + 3\kappa_s} \qquad (26)$$

This simplification, being analogous to the Fick's law of diffusion, is much simpler to solve. However, its validity was observed to be quite limited. In the preliminary simulations, the P1 approximation gave a realistic radiation profile in an optically thick reaction medium. However, in air it was observed to have a uniform spectral irradiance, which does not represent the real case of the inverse square law. Therefore, for the air medium, a different equation had to be considered.

The empirical formula used by Pareek *et al.*[15] to model a linear light source in air was shown to be accurate even as close as 30 mm to the lamp. This equation can be given by

$$G_\lambda = \frac{K_\lambda}{4\pi r}\left(\arctan\left(\frac{z+L}{r}\right) - \arctan\left(\frac{z-L}{r}\right)\right) \qquad (27)$$

where K_λ is the output power of lamp ($\mathrm{W\,m^{-1}}$), L is the length of the lamp (0.45 m), r is the radial distance from the lamp axis (m) and z is the distance from the center of the lamp (m). This empirical formula was implemented into COMSOL by the aid of the wall distance node in mathematics plug-in. r was implemented by calculating the distance from the lamp wall and adding the radius of the lamp (9.5 mm) to obtain the distance from the lamp axis. z was calculated by creating a 0.1 mm thick artificial wall at the mid-point of the lamp as shown in Figure 5.7(b). In other words, the distance of any position in the air domain from the mid-point wall was registered as z and the distance of this position from the lamp axis was registered as r.

The LP K_λ was used as the fitting parameter since the lamp supplier does not supply the net output power for every wavelength band. Since equation (27) is used to calculate irradiance in the air medium, the model shall give accurate irradiance data at the reactor walls using this equation. To ensure correct values at the wall, K_λ values were modified to fit the model output to the G_λ measurements of the empty reactor. The lamp has two different power outputs for the two distinct absorbed peaks shown in Figure 5.4. Two fitted K_λ values were required for these distinct output bands. The fitted values were $K_{220-240} = 0.96\,\mathrm{W\,m^{-1}}$ and $K_{240-270} = 1.30\,\mathrm{W\,m^{-1}}$. With these inputs, the scalar quantity of the irradiance could be calculated at the reactor walls.

Equation (27) is less complicated than the RTE, however, the two different governing equations have different inputs and outputs. Equation (27) has the LP as well as r and z as input and spectral irradiance (G_λ) as output. The output of equation (27) at the reactor walls was used as the input of equation (25). However, equation (25) requires the input of the irradiance source $j_\lambda(\Omega)$, which is a vector quantity at the reactor walls. In addition to the radiation source vector, absorption and scattering constants were also needed as input. To overcome the reactor wall boundary condition problem, calculated G_λ values (by equation (27)) were divided by the solid angle to convert to $j_\lambda(\Omega)$.

For the geometry and case considered, the spectral irradiance was divided to 2π steradian since this is a half-sphere solid angle. This was thought to be valid for the bottom of the reactor since there is no radiation coming from the top and all the G_λ can be directed toward the inner domain of the reactor. On the other hand, the vertical walls were assumed to have a spherical radiance solid angle of 4π steradian since there may be light rays, which exit the reactor sides too. These boundary conditions are schematically shown in Figure 5.7(b).

The scattering constant was taken as zero since the model was developed to calculate irradiance distribution in the beginning of the reaction when there are no solids. Absorption constants were measured by a UV–Vis spectrometer. Two absorption constants for two output peaks of the LPML were required. The absorptivity of the shorter wavelength band was measured as $106 \, \mathrm{m}^{-1}$ and the longer wavelength band was $59 \, \mathrm{m}^{-1}$.

5.5 Results and Discussions

Model validation was done using the setup depicted in Figure 5.6. The reaction medium depth was varied from 5 mm to 30 mm with increments of 5 mm. The LP fitting parameter was tuned with the irradiance measurements taken with the empty reactor. Afterward, irradiance was measured at the fluid surface using the spectrometer. The model results and measurements are shown in Figure 5.8.

Despite the simplified governing equations in both air and liquid, it can be seen that the model follows the measurements with reasonable accuracy. It can also be observed that the model is more accurate with shorter wavelength irradiances. This is due to the P1 approximation. As explained in the model section, the P1 approximation gave unrealistic results in the air medium since there is no light absorption within the wavelength band considered in air. This equation is observed to become more accurate for domains with higher absorptivity. A mesh refinement analysis was also performed to ensure the model results are independent of mesh refinement.

With the confirmed model, the LVREA could be calculated. The LVREA of a full reactor is shown in Figure 5.9(a). The total energy

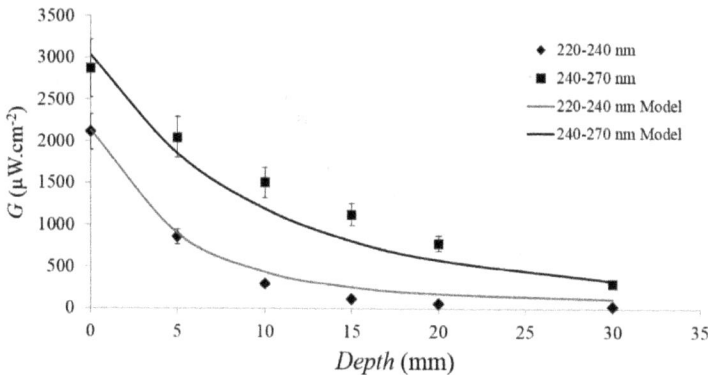

Fig. 5.8. Average irradiance model versus measurements within the shallow batch reactor at different heights. (Reprinted with permission from Ref. 39. Copyright © 2016 Elsevier B.V.)

absorbed by the medium (VREA) for different reaction medium depths is shown in Figure 5.9(b).

Figure 5.9 shows that with increasing reaction medium depth, the absorbed energy density decreases rapidly for the case considered. Figures 5.9(a) and 5.9(b) also show that the increase in total energy absorption is significant till 10 mm characteristic length. Therefore, if a millichannel reactor was to be designed for this light source and reaction medium, it would be significantly more productive (higher STY) if its light path is lower than 10 mm. However, in this case the reaction medium will waste around 0.2 W of power that a longer light-path geometry instead would absorb. Therefore, it can be concluded that, if there are no dominant reverse reactions, a deeper geometry should be chosen to achieve higher efficiency (PSTY). The shallow batch experiments were performed in order to visualize this evolution of efficiency with respect to productivity.

Figure 5.10 shows the Eu concentration decrease in both the shallow batch experiments (0.5–30 mm) as well as deep batch experiments (40 mm and 76 mm) with respect to time.

It can be seen from Figure 5.10(a) that the 5 mm deep reaction medium has a higher reaction rate. Particularly in the first 30 min, before the solid concentration is high, the difference is more obvious.

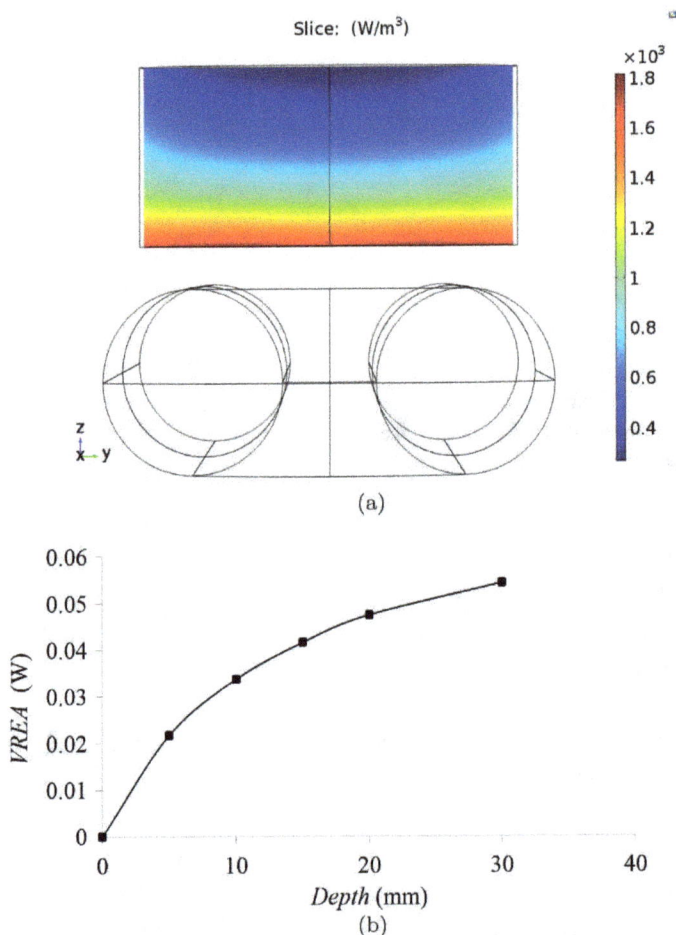

Fig. 5.9. The LVREA distribution within the reactor (220–240 nm band) (a) and VREA with different reaction medium depths (both bands) (b). (Reprinted with permission from Ref. 39. Copyright © 2016 Elsevier B.V.)

In fact, due to the limitations in mixing, the reaction rate of the 5 mm run gets lower than the other runs especially after 50 min. This is due to the geometry of the shallow reactor. At 5 mm depth, the agitator running at 150 rpm could not create the vortices required for efficient mixing and, as a result, settling of the solids was observed. Figure 5.10(b) also shows that changing the irradiance arriving at

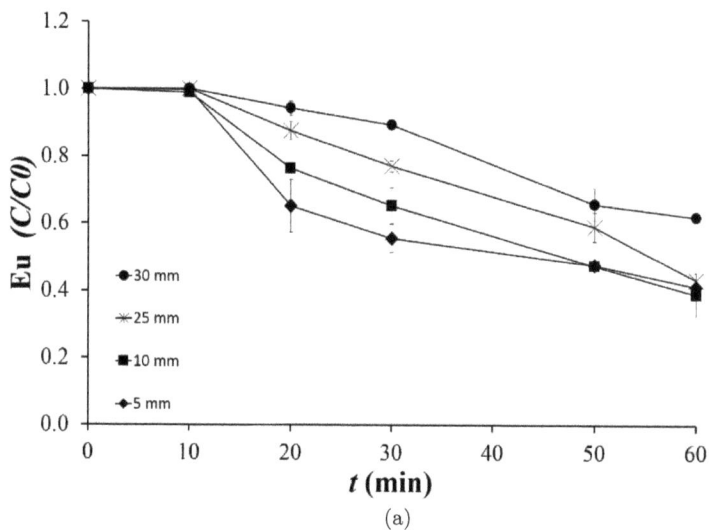

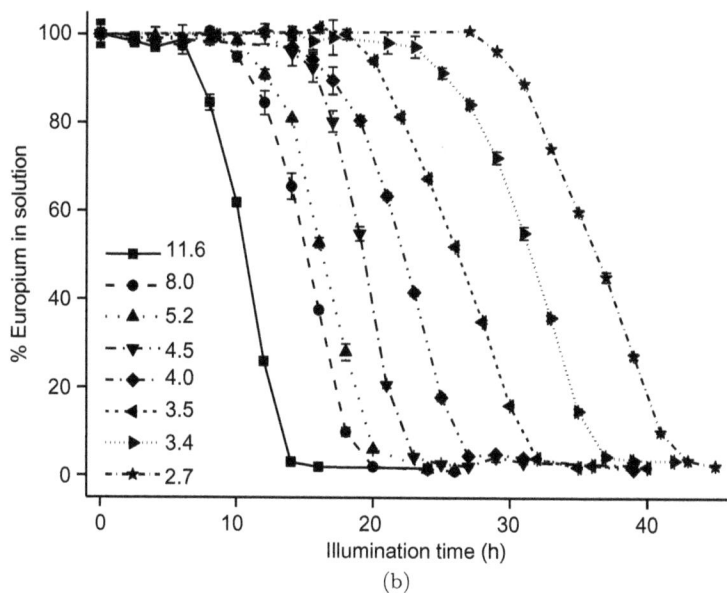

Fig. 5.10. Normalized Eu concentrations versus time for the shallow batch reactor using different reactor depths (a) and for the deep batch reactor using different irradiances (b). ((a) Reprinted with permission from Ref. 39. Copyright © 2016 Elsevier B.V.; (b) Reprinted with permission from Ref. 36. Copyright © 2016 RSC Publishing.)

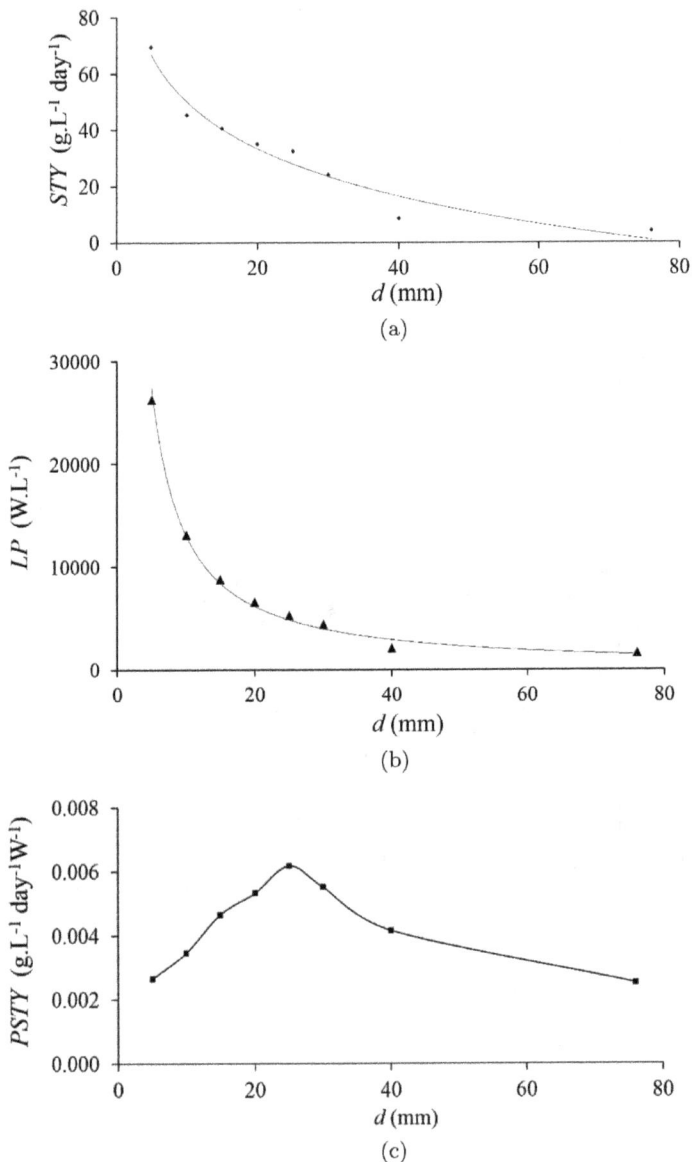

Fig. 5.11. The STY (a), LP (b) and PSTY (c) values with respect to light-path length. The data points of STY and PSTY corresponding to 40 and 76 mm depths are calculated from deep batch experiments. ((a) and (c) Reprinted with permission from Ref. 36. Copyright © 2016 RSC Publishing; (b) Reprinted with permission from Ref. 39. Copyright © 2016 Elsevier B.V.)

the reactor by a factor of 4 has a relatively minor effect in the reaction rate (2-fold change). This indicates that geometry has a more pronounced effect on the productivity than the irradiance at the reactor wall.

To calculate STY and PSTY, the highest production rate obtained in any geometry was utilized. Figure 5.11 shows the interaction of productivity and efficiency with respect to light-path length.

By the aid of Figure 5.11(a), it can be seen that, as expected, the productivity decreases as the light-path through reaction medium gets longer. However, the efficiency (Figure 5.11(c)) increases until the length is 25 mm, which was also expected as explained earlier this section. It can be concluded that this characteristic length of reaction medium provides the optimum balance of productivity and efficiency. As shown in Figure 5.11(b), for a thinner reaction medium, the LP parameter gets high very rapidly due to decreasing reactor volume. The optimum point is reached, since for this case, the STY does not decrease as rapidly as the LP until the optimum point of 25 mm. However, at the right-hand side of the optimum, the LP reaches a plateau and, therefore, the decrease in STY affects the efficiency of the reactor more visibly.

5.6 Conclusion

When the literature of photochemistry and photoreactors is revised, a clear tendency towards micro/millichannel reactors can be observed. This effort is not without a scientific backing. The irradiance within a reactor can decrease within a few millimeters due to the nature of light obeying the rule of the inverse-square law as well as due to the absorbing reaction medium, which obeys the Beer–Lambert–Bouguer law. Therefore, the reactor design has to bring all reaction medium as close as possible to the light source to maximize the LVREA. This necessity is the reason why flow and photochemistry have been so interconnected, and smaller geometries are significantly more productive than just a lamp and a stirred vessel in a batch setup.

Within the scope of this work, the question "How micro shall we go?" was addressed. This was done via the STY and PSTY and it was proven that there is an optimization problem to balance between these two parameters.

It can be concluded that if the reactor design is desired to operate with the highest forward reaction rate and scale-up is not necessary, the reactor geometry can only focus on maximizing the STY. This can be done by using the smallest geometry allowed by the physics of the reaction medium such as viscosity, presence of solids, etc.

However, if scale-up is necessary, the energy efficiency of the photoreactor will have to be taken into account. This is particularly important if a scale-out will be performed, since the efficiencies will add up when the identical reactor units are replicated for the desired capacity increase. Furthermore, the smallest reactor diameter may also cause prohibitive bundling numbers while scaling-out. Leblebici et al.[39] have shown that for a mass-transfer case by opting for a millichannel instead of a microchannel, less numbering-up was necessary for the same industrial capacity. Therefore, compromising a part of the STY would result in a higher energy efficiency as well as a lower amount of reactor and lamp replications.

This phenomenon was shown in this work via a case study on the photochemical reduction of europium(III). For this case, increasing the light-path length of the geometry and the reactor volume five times, decreased the STY by 40%. This may look like a serious compromise, however, it should be noted that the overall product throughput has doubled. Furthermore, the PSTY increased three times, which results in less lamps or weaker lamps to be required for the same throughput. This strategy of photoreactor design has the potential to decrease the cost of both operation and commissioning and can help boost the popularity of photoreactors in industrial applications.

References

1. D. Cambié, C. Bottecchia, N. J. W. Straathof, V. Hessel and T. Noël, *Chem. Rev.* **116**, 10276 (2016).
2. M. E. Leblebici, J. Rongé, J. A. Martens, G. D. Stefanidis and T. Van Gerven, *Chem. Eng. J.* **264**, 962 (2015).

3. Y. Su, N. J. W. Straathof, V. Hessel and T. Noël, *Chem.-Eur. J.* **20**(34), 10562 (2014).
4. O. Carp, *Prog. Solid State Chem.* **32**(1–2), 33 (2004).
5. R. Lakerveld, G. Sturm, A. Stankiewicz and G. Stefanidis, *Curr. Opin. Chem. Eng.* **5**, 37 (2014).
6. T. Van Gerven, G. Mul, J. Moulijn and A. Stankiewicz, *Chem. Eng. Process. Process Intensif.* **46**(9), 781 (2007).
7. A. Visan, D. Rafieian, W. Ogieglo and R. G. H. Lammertink, *Appl. Catal. B Environ.* **150–151**, 93 (2014).
8. M. T. Kreutzer, F. Kapteijn, J. A. Moulijn and J. J. Heiszwolf, *Chem. Eng. Sci.* **60**(22), 5895 (2005).
9. M. E. Leblebici, G. D. Stefanidis and T. Van Gerven, *Chem. Eng. Process. Process Intensif.* **97**, 106 (2015).
10. B. Van den Bogaert, D. Havaux, K. Binnemans and T. Van Gerven, *Green Chem.* **17**(4), 2180 (2015).
11. M. J. Benotti, B. D. Stanford, E. C. Wert and S. A. Snyder, *Water Res.* **43**(6), 1513 (2009).
12. S. Kuhn, T. Noël, L. Gu, P. L. Heider and K. F. Jensen, *Lab Chip* **11**(15), 2488 (2011).
13. O. M. Alfano, I. Cabrera and A. E. Cassano, *J. Catal.* **379**, 370 (1997).
14. M. Motegh, J. Cen, P. W. Appel, J. R. van Ommen and M. T. Kreutzer, *Chem. Eng. J.* **207–208**, 607 (2012).
15. Y. Boyjoo, M. Ang and V. Pareek, *Chem. Eng. Sci.* **111**, 266 (2014).
16. T. G. Goonan, *Rare Earth Elements — End Use and Recyclability*, U.S. Geological Survey Scientific Report, 2011.
17. D. Bauer, D. Diamond, J. Li, D. Sandalow, P. Telleen and B. Wanner, U.S. Department of Energy (2010).
18. Working Group, Ad Hoc, Critical Raw Materials: European Commission, 2014.
19. K. Binnemans, P. T. Jones, B. Blanpain, T. Van Gerven, Y. Yang, A. Walton and M. Buchert, *J. Clean. Prod.* **51**, 1 (2013).
20. K. Binnemans and P. T. Jones, *J. Rare Earths* **32**(3), 195 (2014).
21. D. F. Peppard, G. W. Mason, J. L. Maier and W. J. Driscoll, *J. Inorg. Nucl. Chem.* **4**(5–6), 334 (1957).
22. K. Gupta and N. Krishnamurty, *Int. Mater. Rev.* **37**(1), 197 (1992).
23. S. Cotton, *Lanthanide and Actinide Chemistry* (Wiley, West Sussex, 2005).
24. J. S. Preston and A. C. Du Preez, *J. Chem. Technol. Biotechnol.* **60**(3), 317 (1994).
25. A. G. Atanasyants, A. N. Seregin and A. V. Danilov, *Hydrometallurgy* **44**, 255 (1997).
26. C. A. Morais and V. S. T. Ciminelli, *Sep. Sci. Technol.* **37**(14), 3305 (2002).
27. T. Donohue, *Chem. Phys. Lett.* **48**(1), 2 (1977).
28. T. Donohue, *Opt. Eng.* **18**(2), 181 (1979).
29. L.-F. Qiu, X.-H. Kang and T. S. Wang, *Sep. Sci. Technol.* **26**(2), 199 (1991).

30. H. Hein, H. Hinz, P. Merlet, P. Vetter and U. Bergmann, *Handbuch der Anorganischen Chemie*, 8th edn. (Springer-Verlag, Berlin, 1981).
31. Y. Haas, G. Stein and M. Tomkiewicz, *J. Phys. Chem.* **74**(12), 2558 (1970).
32. S. Muralidharan and J. H. Espenson, *Inorg. Chem.* **23**(6), 636 (1984).
33. A. Matsumoto and N. Azuma, *J. Phys. Chem.* **92**(7), 1830 (1988).
34. V. R. Sastri, J. R. Perumareddi, V. R. Rao, G. V. S. Rayudu and J. C. G. Bünzli, *Modern Aspects of Rare Earths and their Complexes* (Elsevier Science, 2003).
35. C. A. Morais and V. S. T. Ciminelli, *Hydrometallurgy* **60**(3), 247 (2001).
36. B. Van den Bogaert, L. Van Meerbeeck, K. Binnemans and T. Van Gerven, *Green Chem.* **18**, 4198–4204 (2016).
37. M. F. Modest, *Radiative Heat Transfer*, 2nd edn. (Academic Press, San Diego, California, 2003).
38. G. Sgalari, G. Camera-Roda and F. Santarelli, *Int. Commun. Heat Mass Transf.* **25**(5), 651 (1998).
39. M. E. Leblebici, S. Kuhn, G. D. Stefanidis and T. Van Gerven, *Chem. Eng. J.* **293**, 273 (2016).

Chapter 6

Homogeneous Photoreactions in Continuous Flow

Yuchao Deng* and Xiao Wang*,†,‡,§

CAS Key Lab of Low-Carbon Conversion Science and Engineering
Shanghai Advanced Research Institute
Chinese Academy of Sciences 100 Haike Road
Pudong, Shanghai 201210, P. R. China

†*Harvard Neuro Discovery Center*
Harvard Medical School and Brigham & Womens Hospital
Cambridge, MA 02139, USA
‡*wangxiao@sari.ac.cn*
§*xwang21@bics.bwh.harvard.edu*

Her eyes in heaven
Would through the airy region stream so bright
That birds would sing and think it were not night.

— *Romeo and Juliet*, 2.2.20–3

6.1 Introduction

Photons are traceless reagents that brings up amazing changes to a chemical reactivity of organic molecules.[1-3] Although the photoreaction is considered a powerful tool to construct demanding molecular structures in an often energy-saving and sustainable manner, the fear of using photoreactions still exist, mainly in the following aspects: (i) Difficulty of scalability that is due to increasing path length of light transmission according to the Lambert–Beer law. (ii) To reach high conversion, prolonged reaction time is required, which also causes over-irradiation thus side product. These problems are the main obstacles for the widely application of industrial relevant, homogeneous photoreactions.

To address these issues, it is not hard for one to imagine how flow chemistry can possibly improve the efficiency of photoreactions.[4,5] First of all, the enhanced surface-to-volume ratio in a flow reactor as compared to batch would make light irradiation more homogeneously, and significantly enhance the reaction rate.[6,7] Second, the continuous removal of stream mainly consisting of the desired product would reduce the undesired over-irradiation.[8] The early miniature of the photoreactions in continuous flow was first reported by Birr *et al.* in 1972, on the removal of a photolabile protecting group.[9] However, this discovery was buried in the sea of literature until it was rediscovered in recent years. In 2005, Booker-Milburn *et al.* introduced the modern photoflow setup consisting of a medium-pressure Hg lamp, a cooling well and fluorinated ethylene propylene (FEP) tubing, and successfully used the new system to run an alkyne–alkene [2+2] addition and an intramolecular [5+2] reaction.[10] Since then, a number of research groups have sought to find more efficient setups, and more diverse photochemical reactions that well fit in this theme.

This chapter will cover recent development of the homogeneous photoflow reactions, primarily in the past two decades. Examples will be discussed in the different reaction categories.

6.2 Halogenation

One of the commonly applied photochemical transformations is halogenation, because many of them require radicals to initiate. Among all halogenations, bromination was the most investigated one. Časar reported a photochemical bromination of a 5-methylpyrimidine precursor to synthesize Rosuvastatin, as the first example of benzylic bromination in flow (Scheme 6.1).[11] The reactor was constructed using FEP tubing (ID = 0.8 mm) and quartz cooling jacket support to hold a medium-pressure Hg lamp (150 W). With the same yield as

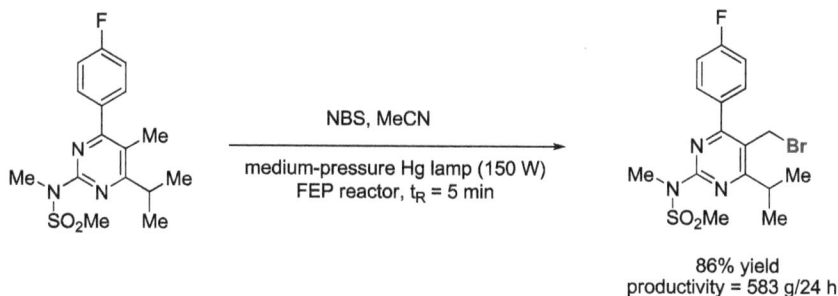

Scheme 6.1. Časar's benzylic photobromination to synthesize Rosuvastatin precursor.

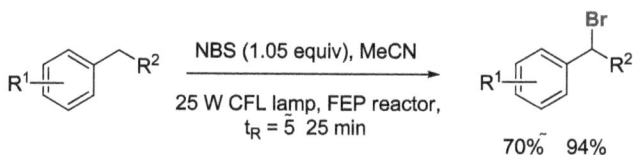

Scheme 6.2. Benzylic bromination in flow.

batch (88%), the flow bromination (with residence time of 5 min) had a superior productivity of 58.3 mmol/h, as compared to the reaction done in a small batch reactor (8.1 mmol/h). The flow bromination offered additional advantages such as purer isolated product (93% purity versus 82% in batch).

Mateos and Kappe also developed a flow method for a facile benzylic bromination with NBS, almost quantitative yields could be realized using a photoflow reactor consisting of FEP tubing and household CFL lamp (23 W) (Scheme 6.2).[12] Temperature probe and cooling fan were installed to monitor and minimize the influence of the heat generated from the bulb. Both small (30-W white CFL or 25-W blacklight) and large-scale (100 W CFL) reactors were tested, and the flow reaction could be scaled up easily. For the reaction with phenylacetone, a high productivity of 180 mmol/h of the mono-brominated product could be realized.

Fluorination of organic molecules provides unique change in their bioactivity and physicochemical properties, and has become an increasing theme for recent research in chemical methodology development. Based on previous studies that benzylic C–H bond can be fluorinated with photocatalyst, Kappe presented a continuous-flow light-induced benzylic fluorinations with residence time of only a few minutes.[13] The batch method ensured excellent isolated yields with a 25-W blacklight CFL as irradiation source, Selectfluor as fluorine source and the economical and practical xanthone as photocatalyst. When adopted in flow, a 105-W CFL blacklight lamp was placed in a glass cylinder coiled with FEP tubing. Compared to the batch method, the flow reactor greatly shortened the reaction time (Table 6.1, Entry 1) from hours to 28 min.

Encouraged by the good yield and selectivity obtained in the substrate scope study, the authors explored reactions on two biologically interesting molecules which may have more challenging selectivity issues (Figure 6.1).[13] Fortunately, 90% fluorination took place in the desired position for the F-ibuprofen methyl ester with an isolated yield of 80%. The fluorination of celestolide also gave a high yield of 87% on a multigram scale.

6.2.1 *Reduction*

The reduction of carbon–halide bond is naturally the reverse transformation of the C–H halogenation, with the addition of a reductant and/or a hydrogen atom source. Stephenson used *fac*-Ir(ppy)$_3$ as the visible-light photoredox catalyst to convert a diverse set of unactivated alkyl, alkenyl and aryl C–I bond to the corresponding C–H bond, with either the Hantzsch ester or formic acid as hydrogen atom donor.[14] The reaction displayed decent yields and excellent tolerance of functional groups (Scheme 6.3). The conventional reagents (such as tributyltin hydride, AIBN, peroxides) for the radical reduction of C–I bond are often associated with safety issues as they are either toxic or explosive, while other SET reductants such as SmI$_2$ suffer air-stability problems. In Stephenson's flow setting, the reaction time was further reduced from 30 h (batch) to merely 40 min. Gratifyingly, only one twentieth of the catalyst loading in batch (1 mol%) was needed in flow (0.05 mol%) to reach comparable yields in a scale-up flow reaction.

Table 6.1. The fluorination of benzylic compounds with a continuous-flow reactor.

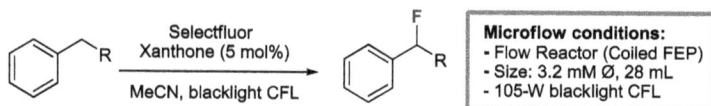

Entry	Products	Temperature (°C)	Conv. (%)
1		25	85[a] 92[b]
2		25	60
3		25	85
4		40	89
5		40	73
6		60	65
7		60	81
8		60	80

(*Continued*)

Table 6.1. (*Continued*)

Entry	Products	Temperature (°C)	Conv. (%)
9		60	71
10		60	73
11		60	69

Note: [a] Continuous flow: residence time = 28 min.
[b] Batch conditions: 25-W blacklight lamp, reaction time = 60 min.

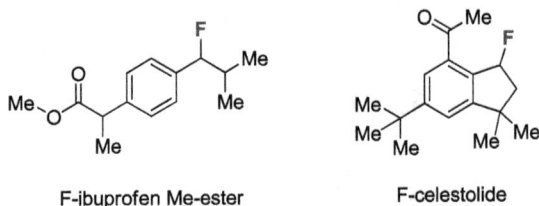

F-ibuprofen Me-ester   F-celestolide

Fig. 6.1. Upscalable flow fluorination of biologically active compounds.

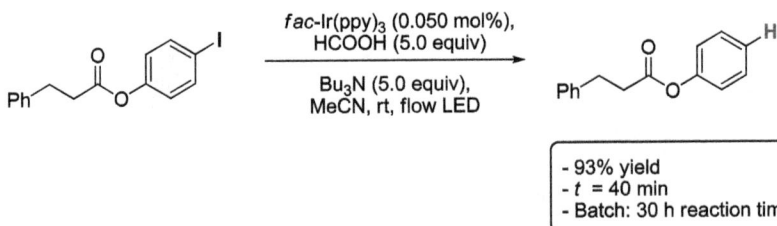

fac-Ir(ppy)$_3$ (0.050 mol%),
HCOOH (5.0 equiv)

Bu$_3$N (5.0 equiv),
MeCN, rt, flow LED

- 93% yield
- t = 40 min
- Batch: 30 h reaction time

Scheme 6.3. Stephenson's continuous-flow reduction of aryl carbon–iodine bond.

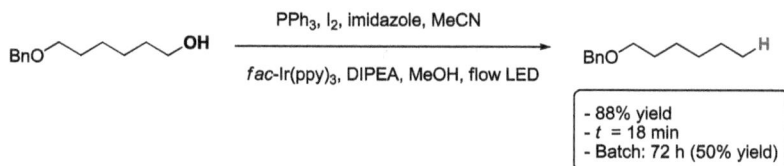

Scheme 6.4. Stephenson's deoxygenation of alcohol in flow.

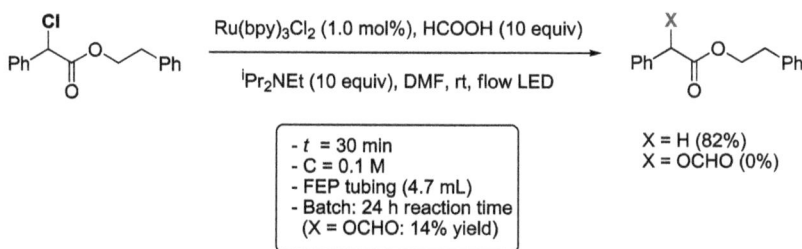

Scheme 6.5. Seeberger's flow reduction of alkyl carbon–chlorine bond.

Motivated by the successful C–I reduction, Stephenson developed a deoxygenation reaction to synthesize alkanes from alcohols (Scheme 6.4).[15] The reaction was operated under milder conditions as compared to the conventional Barton–McCombie conditions. Instead of Hantzsch ester, DIEPA (N, N-diisopropylethylamine) with methanol was used as hydrogen source to increase the solubility of the reaction mixture in the photoflow reactor. Compared to the batch reaction, a 120-fold improvement in productivity was observed in flow.

In general, the reduction of C–Cl bond is more difficult than that of C–I bond. A similar reduction protocol was established by the Seeberger group, in which α-chlorophenylacetates could be reduced in flow reactor with Ru(bpy)$_3$Cl$_2$ as photocatalyst (Scheme 6.5).[16] The FEP tubing was wrapped around two metal rods, which were placed between two cold white 17-W LED lamps. The reduction product was produced in 82% yield in 30 min residence time, while a batch reaction usually took 24 h to complete. What made the flow methods more superior was that the formate side product derived from the

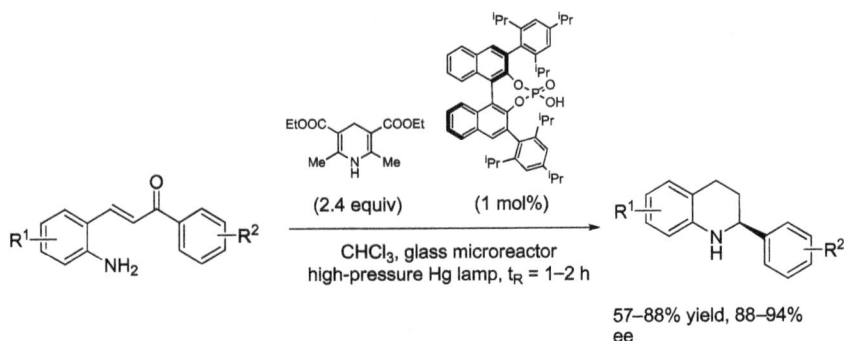

Scheme 6.6. Rueping's photoflow enantioselective cyclization-transfer hydrogenation cascade.

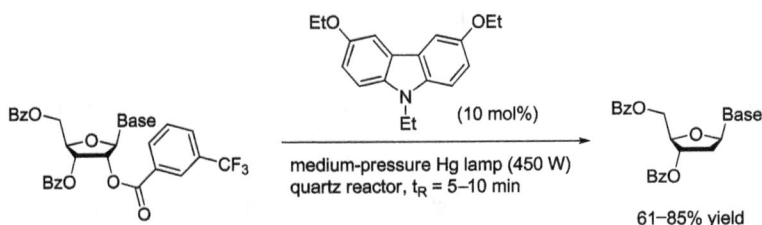

Scheme 6.7. Jamison's photoinduced electron-transfer deoxygenation of nucleoside 2'-*m*-CF$_3$-benzoate in continuous flow.

hydrolysis of the Vilsmeier–Haack intermediate was not seen in flow reaction, while in batch it was obtained in 14% yield.

Tetrahydroquinoline is an important structural motif that can be found in a number of natural products with interesting biological activities. To access this structure, Rueping developed a continuous organocatalytic photocyclization–transfer hydrogenation cascade reaction (Scheme 6.6).[17] The flow reactor was constructed with several commercial glass reactors in parallel, and placed in a water bath to maintain temperature (55°C). The 2-aminochalcone-type substrate could cyclize upon UV irradiation by a high-pressure mercury lamp (TQ 150, $\lambda = 290 - 580$ nm). In the presence of a chiral BINOL-derived phosphoric acid and the Hantzsch dihydropyridine as hydride donor, the cyclized intermediate (a chiral ion pair) was hydrogenated enantioselectively to afford tetrahydroquinoline in >90% ee.

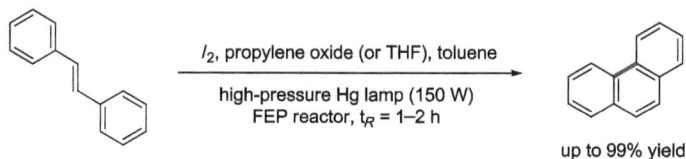

Scheme 6.8. Rueping's photoflow isomerization-oxidative-cyclization of *trans*-stilbene derivatives.

With a quartz tubing photoreactor, Jamison was able to perform photoinduced electron-transfer deoxygenation of nucleoside $2'$-m-CF$_3$-benzoate to prepare $2'$-deoxy and $2', 3'$-dideoxynucleosides, using carbazole as a photosensitizer (Scheme 6.7).[18] A 450-W medium-pressure Hg lamp with a Pyrex sleeve (280 nm cutoff) was used as light source, in combination with a quartz-jacketed immersion well. Quartz reactor tubing (ID $= 1.0$ mm, volume $= 1.84$ mL) was placed around the immersion well, which is cooled by tap water to keep low temperature and ensure good regioselectivity. In addition, UV-reflecting aluminum mirror was placed around the light source to enhance the light intensity, thus promoting the reaction rate (10 min in flow versus 2 h in batch).

6.2.2 Oxidation

The oxidative formation of aryl C–C bond to construct heterocycles is one of the long-pursued themes in organic synthesis.[19] Rueping reported a UV-driven continuous-flow oxidative cyclization of stilbene-type derivatives to synthesize phenanthrenes and helicenes (Scheme 6.8). Iodine molecule was used as the homogeneous oxidant, in the presence of either THF or propylene oxide to quench the subsequently formed HI. To build the flow apparatus, UV-transparent FEP tubing (OD $= 1.0$ mm, ID $= 0.5$ mm, or the one with larger volume of OD/ID $= 4.0/2.0$ mm) was wrapped around the water-cooling unit with a high-pressure Hg lamp (150 W). Up to 99% yield could be obtained for the isomerization-oxidation of *trans*-silbene. Many other phenanthrenes and helicenes were prepared with the photoreactor in moderate to good yields, with residence time of 2 h.[20]

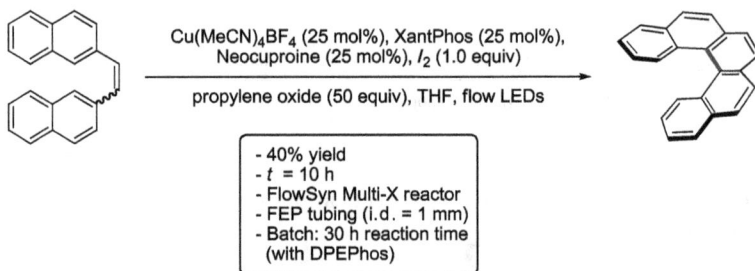

Scheme 6.9. Collins' photocyclization of stilbene derivatives in a microreactor.

Similarly, with a FlowSyn Multi-X reactor equipped with the highly light-transmittable FEP tubing, the Collins group was also able to perform the photocyclization of stilbene-type molecules to access the inherently helically chiral [5]helicenes.[21] The photocatalyst they employed consists of a Cu(I) core modified with XantPhos (or DPEPhos) and neocuproine, while other popular Ru and Ir photoredox catalysts failed to give good yields (Scheme 6.9). The major side product formed by the second oxidative cyclization of the desired product was minimized with the visible-light photocatalysis as compared to the UV method. The residence time of the flow reaction was unusually long (10 h); nevertheless, compared to the batch reaction (120 h), it was significantly faster.

With the same visible-light photosensitizer ([Cu(Xantphos) (dmp)]BF$_4$), Collins' team developed a novel synthesis of carbazoles from diarylamines or triarylamines.[22] With I$_2$ as oxidant and propylene oxide/THF as HI quencher, a number of carbazoles could be made in up to 85% yield in batch. With the flow microreactor (ID = 0.5 mm), a variety of arylamines could be smoothly converted to the corresponding nitrogen-containing heterocycles in yields of 50–80% (Table 6.2, Entries 1–8). The flow reactor had an obviously effect in shortening the reaction time (20 h in flow versus 120 h in batch) and improved the productivity (Table 6.2, Entry 9).

Later, Collins' team compared two different illuminants for the preparation of carbazoles compounds.[23] They tried both visible-light and UV light as irradiation source and synthesize carbazole skeletons from triarylamines, under continuous microflow conditions

Table 6.2. Collins' continuous-flow oxidative photocyclization of arylamines.

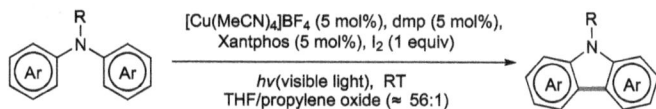

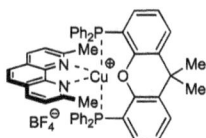

Entry	Substrate	Time (h)	Product	Yield (%)
1		20		65
2		20		79
3		20		65
4		20		64
5		20		51
6		20		53
7		20		60

(*Continued*)

Table 6.2. (*Continued*)

Entry	Substrate	Time (h)	Product	Yield (%)
8		20		63
9		10 2 weeks[a]		75 85[b]

Note: [a]The reaction time in batch.
[b]The yield in batch.

(Scheme 6.10). [Cu(Xantphos)(dmp)]BF$_4$ was again employed as photosensitizer when visible light was used. Most triarylamines with electron-rich substituents had higher yields than electron-poor triarylamines. The visible-light-mediated method with the Cu-based sensitizer sufficiently provided good yield when triarylamines with nitrogen-based heterocycles and halogen-containing arenes. In general, UV-light methods provide shorter residence time than visible-light methods in the same continuous-flow reactor.

6.2.3 *Addition reactions*

In 2005, Booker-Milburn reported for the first time a novel, sustainable, widely applicable flow apparatus wrapped with FEP, for the [2+2] photocycloaddition of maleimide and 1-hexyne (Scheme 6.11(a)).[10] By varying the parameters of the FEP flow reactor, such as layers, flow rate and diameter, maximum output could be achieved. The authors tested multiple reactors eventually found that using a wider bore tubing (ID = 2.7 mm, OD = 3.1 mm) could break through the flow rate limitation of narrow bore FEP tubing (ID = 0.7 mm) and could also avoid the drawbacks of the narrow tubing which was easily blocked and ruptured. The continuous [5 + 2] intramolecular photocycloaddition reaction (Scheme 6.11(b)) demonstrated again the

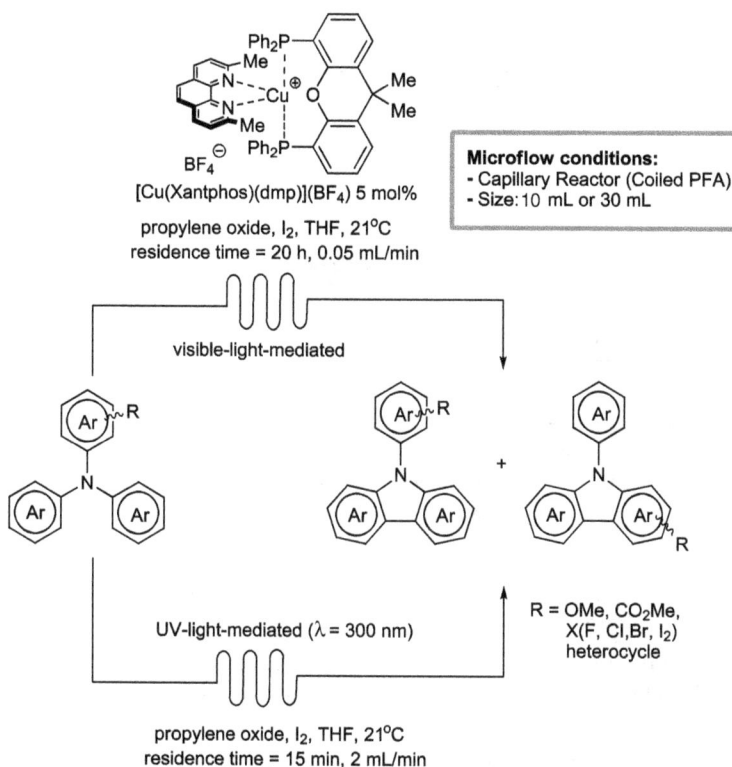

Scheme 6.10. Comparison of UV- and visible-light photocyclizations of triarylamines.

advantage of wide bore tubing and also ensured high isolated yields. Consecutive operation of 24 h with 400-W medium-pressure Hg lamp at 0.1 M (4 mL/min) provided a decent yield of the addition product on a larger scale (85 g, 82% yield).

Booker-Milburn later completed a total synthesis of (±)-neostenine in 14 steps from furan by applying an intramolecular [5 + 2] maleimide photocycloaddition to implement the pyrrolo[1,2-α]azepine core.[24] A moderate yield (40–60%) could be achieved in a small scale (50 mg) reaction in an immersion-well batch photoreactor, however, scaling up to >100 mg scale with the same batch reactor only got less than 20% yield. To improve the results, the

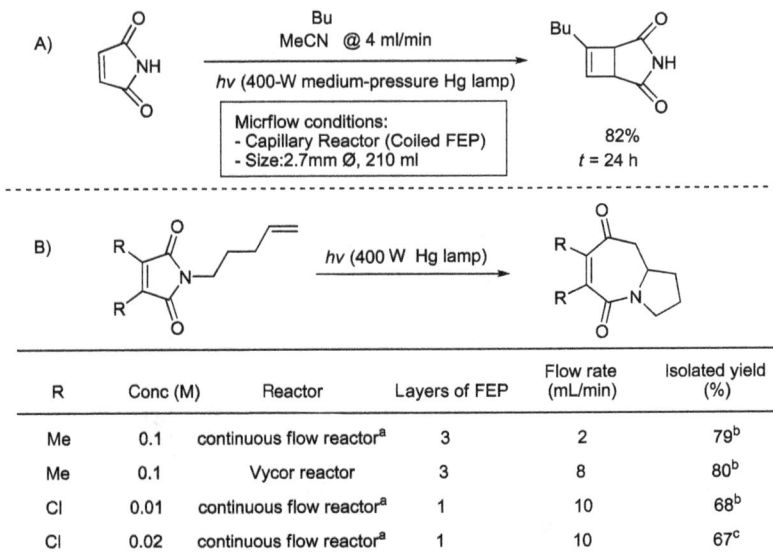

Scheme 6.11. (a) [5 + 2] photocycloaddition in different reactors; (b) [2 + 2] photocycloaddition of maleimide in continuous-flow reactor.

authors performed the reaction in the continuous microflow photoreactor with high-power UV source (400-W medium-pressure Hg lamp) flowing with a diluted solution (0.001 M) at an elevated flow rate (11 mL/min), and desirable yields was reached (63% yield, 1.3 g after 9 h operation), albeit 20% of unreacted reactant was recovered (Scheme 6.12).

For UV-mediated reactions, besides the commonly used conventional high-pressure Hg lamp, Ryu applied low-power blacklight and UV LEDs as alternative, energy-saving light sources for photoreactions.[25] For the photocycloaddition of cyclohexenone with vinyl acetate (Table 6.3), the 1.7-W UV LED achieved the best energy efficiency, and the 15 W blacklight also provided higher energy efficiency than the 300 W Hg lamp (Table 6.3, Type B). Moreover, a satisfactory yield of 91% was observed after just 15 min irradiation with the 1.7-W UV LED. Ryu also compared three different microreactors, in which both the blacklight and UV LED achieved the good results.

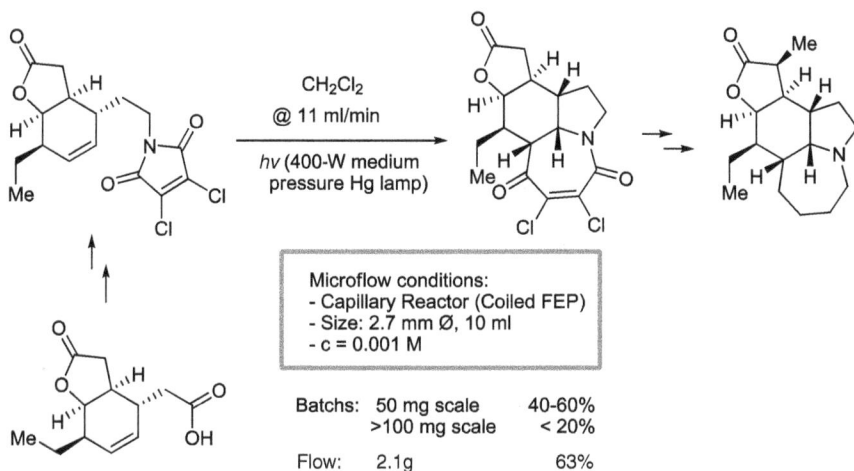

Scheme 6.12. [5 + 2] photocycloaddition of preparation pyrrolo[1,2-α]azepine core.

To test the generality of this photoflow Paterno–Büchi addition in Type A and B microreactors, Ryu tested other ten combinations of cyclohexenones and alkenyl esters, and obtained moderate to good yields with 300-W Hg lamp or 15-W blacklight.

Kakiuchi reported a diastereoselective [2 + 2] cycloaddition of a chiral cyclohexanone with ethylene, and achieved high conversions and diastereoselectivity (53% in Pyrex tube after 5 min; 52% in flow reactor after 1 min irradiation, Scheme 6.13).[26] By injecting ethylene gas and the solution of chiral substrates with syringe pump, they created a slug flow with gas and liquid layers. Upon irradiation by a 500-W high-pressure Hg lamp, 100% conversion could be completed in merely 1 min, as compared to the batch reaction that provided 67% yield in 5 min. The diastereomeric excesses in flow and batch experiments are both moderate (around 50%).

In addition, the slug flow forms a thin layer between the ethylene gas layer and the inner wall of the FEP tubing, in which an even shorter path length of light was achieved (Figure 6.2). This allows for an even better transmission (100%, path length <0.1 mm) compared to that in the normal flow tuning (86%, ID = 1.0 mm). Note that the average transmission in a Pyrex test tube is much lower (12%,

Table 6.3. Ryu's photocycloaddition in three microreactors with different light sources.

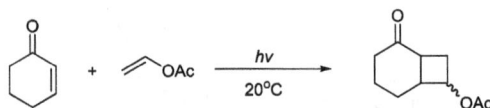

Microreactor	Microflow conditions	hv source	Yield (%)	t	Wh	Yield/ Wh
Type A	— Capillary reactor (Forturan glass)	300-W Hg lamp	88	2 h	600	0.15
	— Size: 1000 μm width; 500 μm depth; 1.9 m length; 0.95 μL volume	15-W blacklight	48	2 h	30	0.16
Type B	— DNS photomicroreactor (quartz plate)	300-W Hg lamp	71	2 h	600	0.12
	— Size: 1000 μm width; 300 μm depth; 2.35 m length; 0.7 mL volume	15-W blacklight	82	2 h	30	2.7
		1.7-W UV LED	91	15 min	0.425	2.14
Type C	— Capillary reactor — Size: 1000 μm width; 200 μm depth; 56 cm length; 0.112 mL volume	1.5-W UV LED	95	20 min	0.5	190

ID = 14 mm) (Figure 6.3). The formation of slug flow results in an increased amount of cyclohexenone in the excited state.

Yoshida demonstrated that a Pauson–Khand reaction could be accelerated by using a photochemical protocol in a microreactor setting.[27] The mixture of the phenylacetylene-dicobalt complex and the norbornene-type substrate was pumped through the tubular quartz reactor, which was irradiated with light emitted from a medium-pressure Hg lamp, so that the cobalt complex is effectively

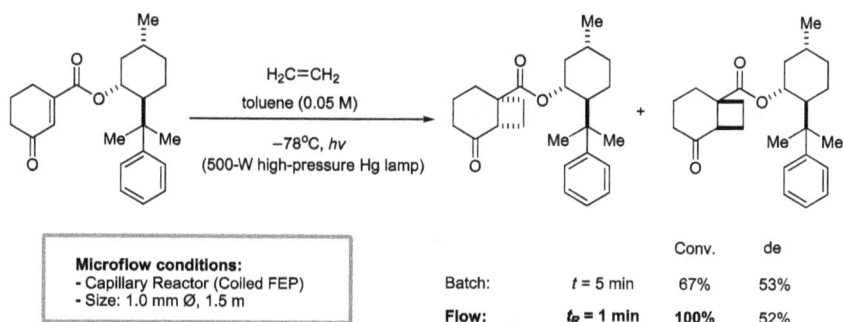

Scheme 6.13. The diastereoselective [2 + 2] photocycloadditions in flow.

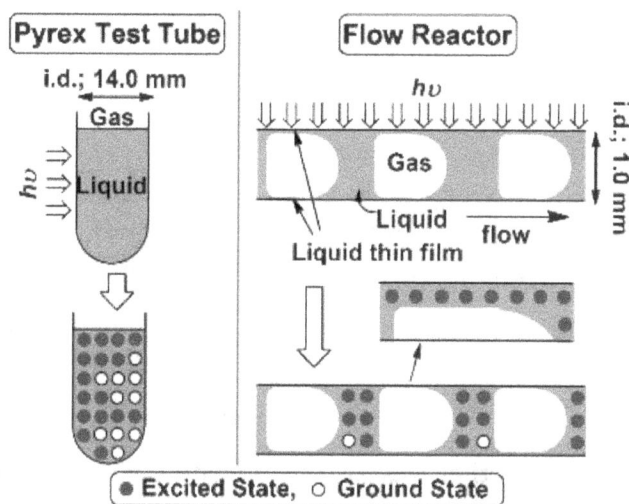

Fig. 6.2. Irradiations in batch and slug flow modes. Copyright © 2012 Akadémiai Kiadó.

activated. The irradiated mixture was collected in a batch reactor in the downstream and continued with the cyclization step of the Pauson–Khand reaction in 5 min (Scheme 6.14). The residence time varied from 55 s to around 600 s, and high conversion of a number of substrates could be realized (yield up to 92%). In contrast, the photobatch reaction without pre-activation of the complex with a photoflow reactor suffered low yield (around 30%). Intramolecular

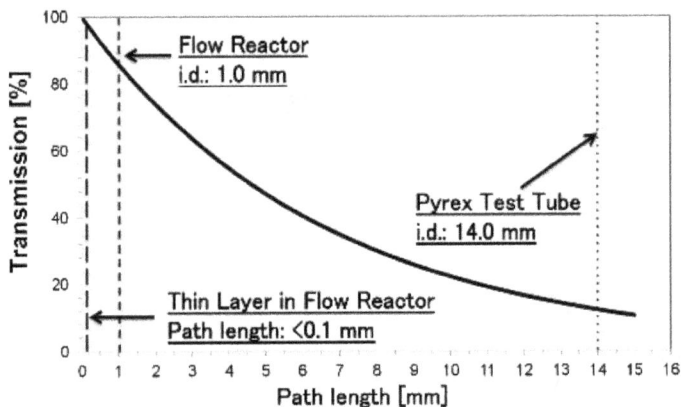

Fig. 6.3.   A comparison of transmission in slug flow, normal flow and test tube. Copyright © 2012 Akadémiai Kiadó.

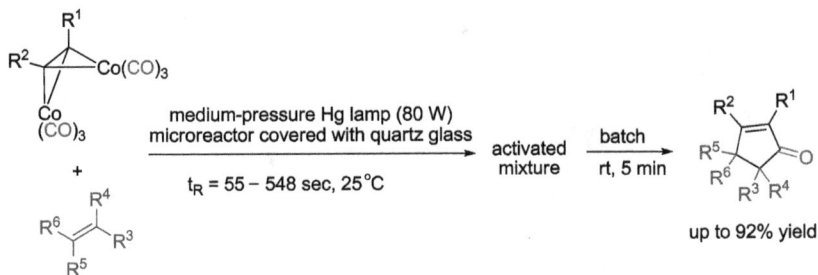

Scheme 6.14.   Activation of Pauson–Khand reaction mixture in flow.

Pauson–Khand reaction was also investigated with the same flow assembly with satisfactory results.

Jamison's group developed a Ru-catalyzed flow version of the ene–yne coupling reactions (Scheme 6.15).[28] The photolysis of $CpRu(\eta^6\text{-}C_6H_6)PF_6$ in flow generated the catalytically active species $[Ru(Cp)]^+$ *in situ*, which catalyzed the coupling of alkene and alkyne to afford a diverse set of skipped diene products with different functional groups, in decent yields and high E/Z ratio. The core of the flow reactor consisted of a 450-W Hg lamp and PFA tubings. Batch reaction took up to 48 h, while the flow version could be completed in merely 5 min. The catalyst, $CpRu(\eta^6\text{-}C_6H_6)PF_6$, could be totally recovered at the downstream of the reactor.

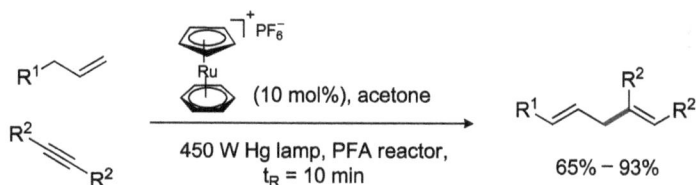

Scheme 6.15. Jamison's Ru-catalyzed ene–yne coupling in flow.

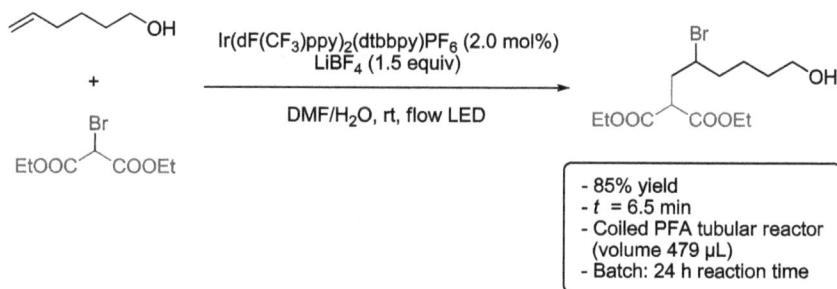

- 85% yield
- *t* = 6.5 min
- Coiled PFA tubular reactor (volume 479 µL)
- Batch: 24 h reaction time

Scheme 6.16. Continuous-flow atom transfer radical addition between alkene and alkyl bromide by Stephenson.

Stephenson was able to optimize the atom transfer radical addition (ATRA) between alkene and alkyl bromide with the same microfluid reactor as of his carbon–halide bond reduction.[29] In the presence of Ir(dF(CF$_3$)ppy)$_2$(dtbbpy)PF$_6$, high yields (85%) and short residence time (6.5 min) could be achieved in the coupling of terminal alkenyl alcohols with alkyl bromides (Scheme 6.16). It was worth noting that the batch experiment of the same substrate typically took 24 h.

Although continuous microflow reactors have demonstrated unique superiorities over the traditional batch reactors in a number of important reaction categories, it also has some disadvantages.[30] One of the major ones is that in handling crystallized, indiscerptible and large protein molecules, the reactor is easily clogged or even ruptured. In order to solve this problem, Horie designed a system that used liquid/gas slug flow and ultrasonic equipment to avoid clogging.[31] The FEP tubing was immersed in the beaker that is placed in an ultrasonic batch. N$_2$ gas was introduced to the system through a

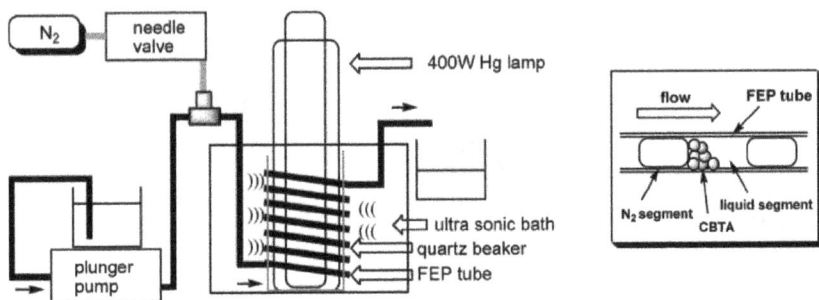

Fig. 6.4. Horie's single-pass microreactor. Copyright © 2010 American Chemical Society.

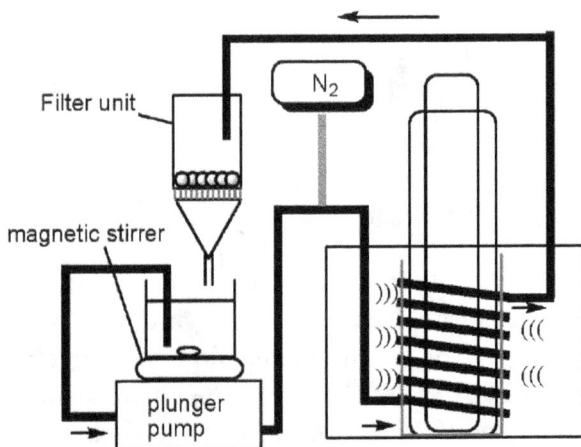

Fig. 6.5. Horie's continuous recycle flow microreactor with filter unit. Copyright © 2010 American Chemical Society.

T-shape mixer. The function of the N_2-segment was to separate the flow into numerous "microbatch reactors" that contains and pushes forward their own precipitate individually (Figure 6.4).

In addition, Horie used continuous recycle flow microreactor to extend residence time, improve the conversion rate of reaction, and reduce the waste of raw materials (Figure 6.5).

As one major application of their reactor design, flow synthesis of cyclobutane tetracarboxylic dianhydride (CBTA) from maleic anhydride was done (Scheme 6.17).[31] The reaction mixture

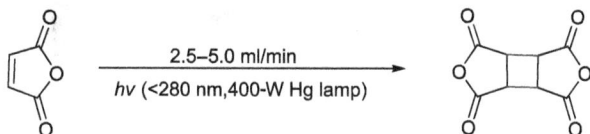

Microflow conditions:	Batch:	t = 6 hours	26%
- Capillary Reactor (Coiled FEP)	Single flow:	rt = 11.6 min	29%
- Size: 1.2 mm Ø, 11.2 m	**Recycle Flow:**	**rt = 11.8 min**	**69% (4.5 cycles)**

Scheme 6.17. Horie's continuous-flow photodimerization of maleic anhydride.

underwent multiple passes through the reactor so that enhanced yield could be realized. Approximately 69% of cycloaddition product was obtained after recycling 4.5 times in 9 h operation (residence time 11.8 min). In a single-pass of flow reaction, only 29% yield could be achieved of after 6 h operation (residence time 11.6 min). More gratifyingly, because of the shortened irradiation time in the flow reactor, the flow reaction suffered less side reactions as compared to the batch equivalent.

Continuing their success in the dual-microcapillary system, Oelgemöller and coworkers developed a new multimicrocapillary flow reactor (MμCFR) for performing parallel synthesis.[32] A 10-syringe pump was used to deliver reactant and reagent solutions to reach a practical number of parallel reactions. Two groups of five FEP tubings were wrapped around two Pyrex tubing fitted with UVA fluorescent tubes (λ_{max} = 365 nm). At the bottom of the two columns there were two cooling fans for maintaining low temperature (Figure 6.6).

This novel device was applied to the synthesis of a series of lactone derivatives via the DMBP (4,4′-dimethoxybenzophenone)-sensitized photoaddition of secondary alcohols to 2(5H)-furanones with different 5-substituents. The authors also compared the MμCFR with batch and several other flow reactors (Table 6.4). High conversions and yields were achieved with the new system with a shorter time (10 min) compared to batch (20 min). Although the reduction of reaction time in the MμCFR was not so significant, the MμCFR consumed only 70% energy of batch reactor and did not require cooling water.

Fig. 6.6. Multimicrocapillary flow reactor. (a) Collection flasks; (b) FEP micro-capillaries; (c) 10-syringe pump. Copyright © 2012 American Chemical Society.

6.2.4 *Cross-coupling*

Cross-coupling reactions under photoflow conditions would be attractive, especially in pharmaceutical industry, however, it has been underdeveloped so far. In 2013, Wang developed a one-pot, photocatalytic version of the Stadler–Ziegler reaction to prepare aryl C–S bond.[33] Upon treatment with organic nitrite, the aniline was converted to diazonium salt *in situ*, which was coupled with thiol to form diazosulfide. In the presence of Ru photocatalyst and visible light from a commercial CFL lamp, the explosive diazosulfide mildly and completely decomposed to afford the desired arylsulfide in good yield (Scheme 6.18). With a microreactor assembly developed by Noël, the efficiency of the reaction could be elevated to a higher level (15 s in flow versus 5 h in batch). The flow assembly consists of a blue LED, PFA tubing (ID = 0.5 mm, volume = 0.45 mL) and aluminum foil to increase photonic efficiency.

Table 6.4. Microreactor of 2(5*H*)-furanones photoadditions.

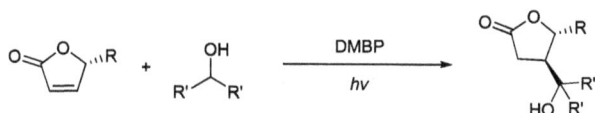

Reactor	Reactor conditions	R	R'	Time (min)	Conv. (%)	Yield (%)
Batch	— Pyrex tubes	H	CH$_3$	20	100	94
	— 16 × 8-W lamps	OEt	CH$_3$	20	100	88
	($\lambda \geq 300$ nm)					
	— Size: 1 cm Φ	OMent	CH$_3$	20	100	87
Dwell device[a]	— UV panel	H	CH$_3$	10	100	—
	— 5 × 8-W	OEt	CH$_3$	10	100	—
	UVA-lamps					
	Size: 0.5 × 2 mm,	OMent	CH$_3$	10	100	—
	1.68 mL					
Microchip	— glass	H	CH$_3$	20	100	94
	— 6 × 75 mW LEDs	OEt	CH$_3$	20	100	88
	— Size	OMent	CH$_3$	20	100	87
	150 × 150 μm,					
	13 μL					
Microcapillary[b]	— Capillary Reactor					
	(Coiled PTFE)					
	— 8 W UVA lamp	H	CH$_3$	10	100	—
	($\lambda = 300 \pm 25$ nm)					
	Size: 0.558 μm $\varnothing$,	OEt	CH$_3$	10	100	—
	1.12 mL					
	— c = 33.3 mM	OMent	CH$_3$	10	100	—
Multimicrocapillary	— Capillary Reactor	H	CH$_3$	10	—	94[c]
	(Coiled PTFE)					
	— 2 × 18 W UVA	OEt	CH$_3$	10	—	60[c]
	fluorescent					
	($\lambda_{\max} = 365$ nm)	OMent	CH$_3$	10	—	90[c]
	Size: 0.8 mm $\varnothing$,	H	C$_2$H$_5$	20	—	80[d]
	— c = 33.3 mM	OEt	C$_2$H$_5$	20	—	57[d]
		OMent	C$_2$H$_5$	20	—	61[d]

Flow rate: [a]168 μL/min, [b]115 μL/min, [c]0.5 mL/min, [d]0.25 mL/min.

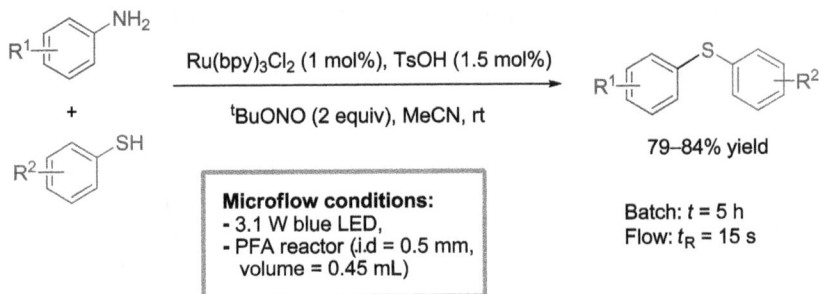

Scheme 6.18. Wang and Noël's photocatalytic Stadler–Ziegler reaction.

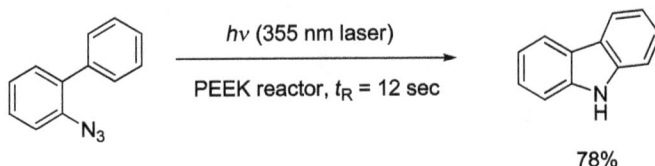

Scheme 6.19. Bräse's synthesis of carbazole by photolysis.

6.2.5 *Insertion*

Aryl azide is often employed as a photolabile precursor of nitrene for a series of insertion reactions. Bräse developed a flow synthesis of carbazole from azidobiphenyl utilizing laser irradiation at 355 nm (frequency-tripled Nd:YAG) (Scheme 6.19).[34] Laser is advantageous comparing to traditional light sources, because it provides a beam of more intensified energy to the irradiated area. To test the efficiency of the laser photoreactor, the authors performed the cyclization of azide, whose absorption lies exact around the emission wavelength (355 nm). The material used to build the reactor was polyether ether ketone (PEEK). Amazingly, the flow reaction could be completed in 12 s in 78% yield, while the batch reaction using Xe lamp (400 W, >345 nm) took 18 h and the yield was much lower (50%).

Starting from aryl azides without the biphenyl structure, Seeberger demonstrated a flow synthesis of 3*H*-azepinones and related azepine derivatives using a flow reactor with FEP capillary wrapped around a 450-W medium-pressure Hg lamp.[35] Within 15–30 min, a

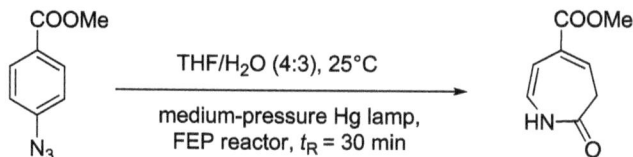

Scheme 6.20. Seeberger's flow photolysis of aryl azide.

variety of products were produced with moderate to good yields (Scheme 6.20).

6.2.6 *Functionalization*

With unique bioactivities, organic molecules with the polyfluoroalkyl chains have wide utilities in pharmaceutical and agrochemical research. In order to simplify the operation and improve the productivity, Kappe developed a novel continuous-flow microreactor for the two-step photocatalytic synthesis of α-trifluoromethyl ketones (Figure 6.7).[36] The novel device contains two tubular reactor units for the TMS-enol ether formation and the trifluoromethylation respectively. The second reactor is made of transparent FEP tubing (ID = 3.18 mm) with a 30-W household CFL lamp as light source.

Phenacyl chloride was identified as the major side product due to the electrophilic addition of the chlorine in CF_3SO_2Cl to the TMS-enol ether intermediate. Kappe chose the cheap, non-toxic organic dye Eosin Y to replace the pricey Ru photocatalyst, which was less effective and selective in batch reaction. The flow reaction apparatus not only increased the selectivity and ensures good conversions (Scheme 6.21, 93% conversion, 86% yield), but also cut down the catalyst loading (0.5 mol% as compared to 1 mol% for batch). In general, excellent yields were obtained with other aromatic and aliphatic substrates with either acyclic or cyclic structures.

Zeitler transplanted MacMillan's organocatalytic, enantioselective α-alkylation of aldehyde into flow.[37] To replace the expensive metal catalyst, organic dye Eosin Y was used which provided comparable results. In a microreactor setting (0.6 mm × 0.5 mm), high yields with much shorter reaction time (typically 45 min as compared

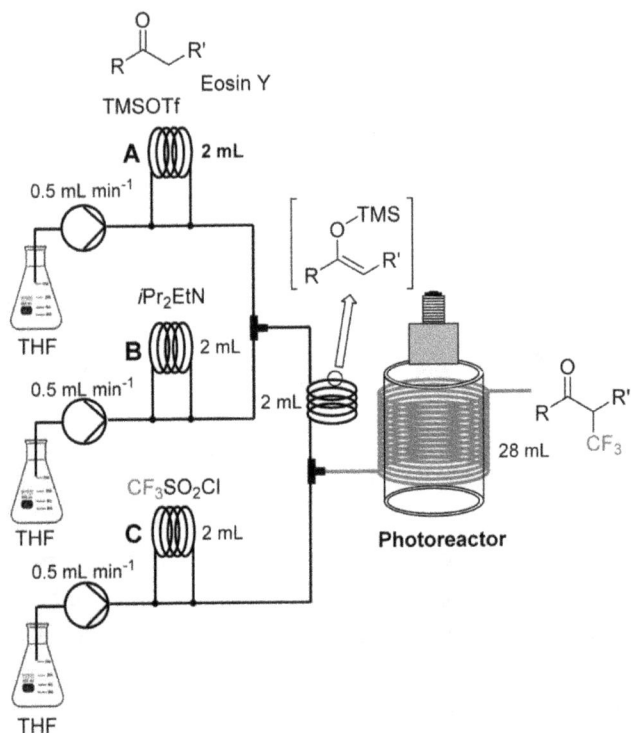

Fig. 6.7. Kappe's continuous-flow setup for the α-trifluoromethylation of ketones. Copyright © 2014 American Chemical Society.

to 18 h in batch) were obtained (Scheme 6.22). The reactor's temperature was maintained at $-5°C$ to ensure good enantioselectivity, which was shown to be the same as the one in batch (ee = 70–90%). For scale-up, an FEP tubing (ID = 0.8 mm) was coiled around a glass beaker, in the center of which a fluorescent bulb (23 W) was placed. The photocatalytic dehalogenation of α-halide carbonyl compounds, and the Stephenson's aza-Henry reaction were also explored with the same reactor design, and high yield could be achieved.

With continuous-flow technology, Stephenson and coworkers optimized several radical cyclization reactions previously reported by them (Scheme 6.23).[29,38] Seven prearranged high power blue LEDs (λ_{max} = 447.5 nm, 5.88 W each) were placed around PFA tubing (ID = 0.762 mm) to ensure maximum absorption. Several 1-pyrrolyl

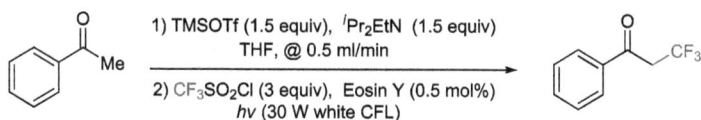

Microflow conditions:
- Capillary Reactor (Coiled FEP)
- Size: 28 ml
- Total flow rate: 1.5 ml/min

Batch: t = 15 min
Flow: t = 20 min

conv (%)
85
93

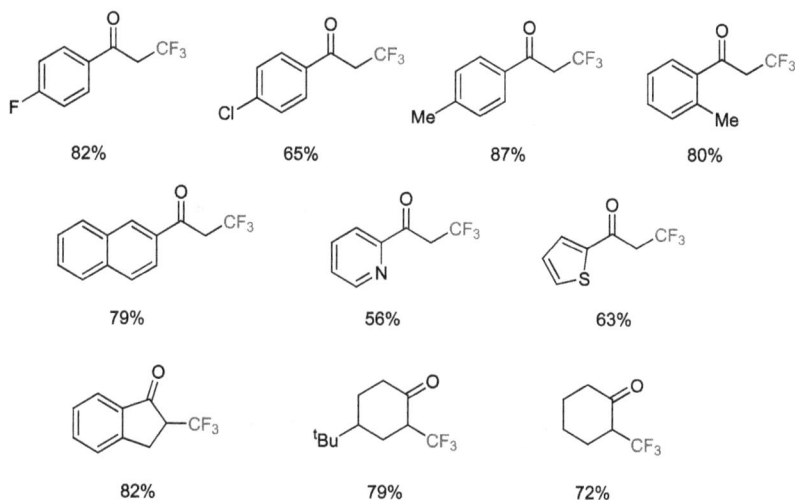

82%   65%   87%   80%

79%   56%   63%

82%   79%   72%

Scheme 6.21. Kappe's α-trifluoromethylation of acetophenone derivatives in flow.

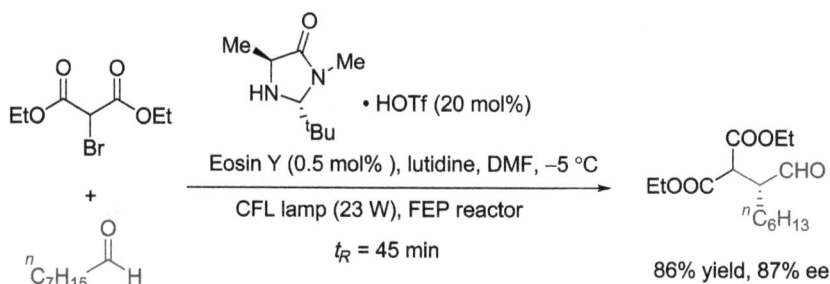

Scheme 6.22. Zeitler's asymmetric α-photofunctionalization of aldehyde in continuous flow.

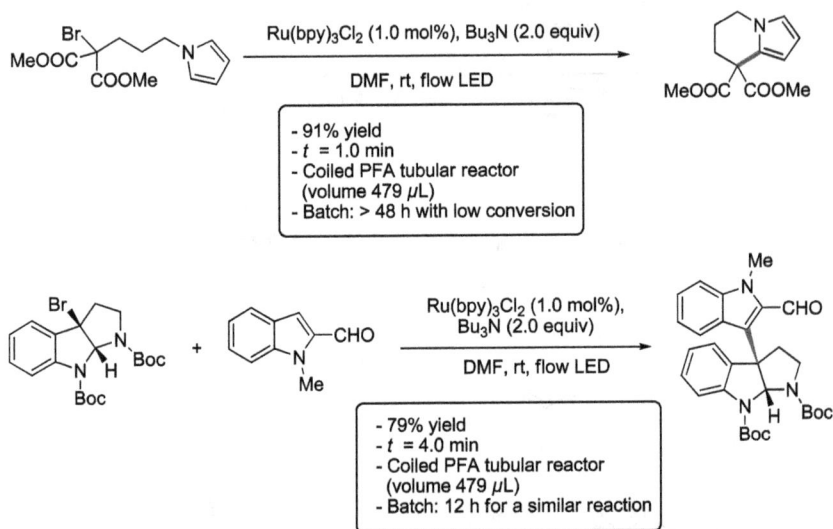

Scheme 6.23. Radical pyrrole and indole functionalization in photoflow reactors.

alkylbromides were tested, and they underwent intramolecular cyclization to the six-membered ring product in 91% yield with only 1 min of residence time. While the batch cyclization suffered low conversion, scaling up with the flow system was quite feasible in gram scale. Using the same flow system, several pyrrole derivatives could be successfully alkylated with a tertiary bromide in only 1–4 min residence time, in up to 80% yield.

The photogenerated iminium ion can also be precursors to make symmetrical anhydrides from carboxylic acids.[39] Similar to the α-functionalization of amine, the Vilsmeier–Haack reagents (bromonium ion) could be formed *in situ* upon visible-light irradiation in the presence of CBr_4. The Vilsmeier–Haack reagent served as an activating agent for the carboxylic acid, thus facilitates the formation of anhydrides. A diverse set of symmetric anhydride could be prepared under mild conditions in Stephenson's photoflow reactor. For example, the anhydride of 4-*tert*-butylbenzoic acid could be synthesized in almost quantitative yield (97%) after passing the flow reactor in 6.4 min (Scheme 6.24).

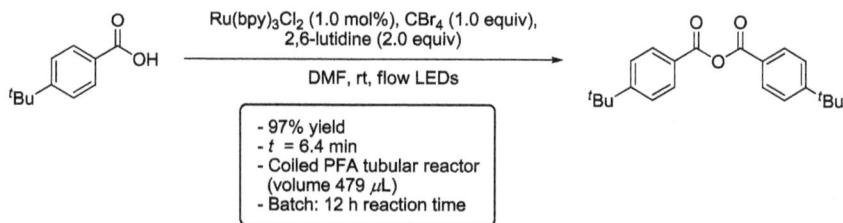

Scheme 6.24. Stephenson's photocatalytic anhydride synthesis in flow.

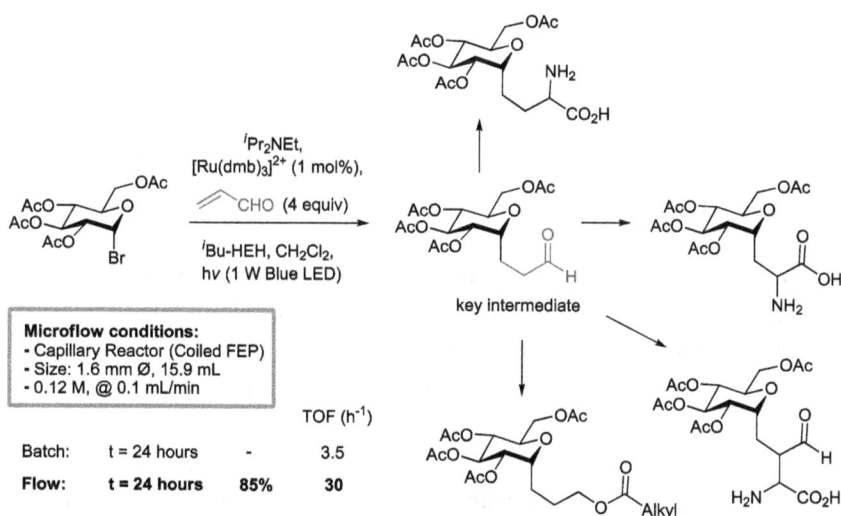

Scheme 6.25. Gagné's synthesis of C-glycoside in continuous-flow reactor.

Glycoconjugates are important biologically active molecules that are widely used in the intercellular recognition vaccine therapeutics. C-Glycolipids with formyl group are key intermediates of the synthesis of C-Glycoconjugate derivatives, since many functionalities such as amino acids and esters can be introduced through the terminal aldehyde (see Scheme 6.25). Previous synthesis all required multiple synthetic steps and scaling up the synthesis remained challenging.[40-42] Gagné and coworkers developed a one-step light-mediated conjugate addition of bromide to acroline catalyzed by [Ru(dmb)$_3$]$^{2+}$. By comparing the yields by using the traditional flask

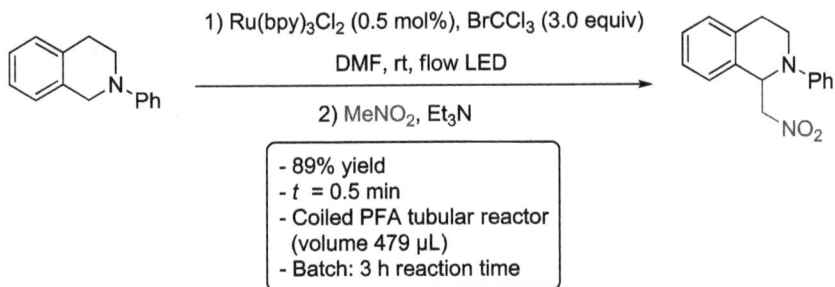

1) Ru(bpy)$_3$Cl$_2$ (0.5 mol%), BrCCl$_3$ (3.0 equiv)

DMF, rt, flow LED

2) MeNO$_2$, Et$_3$N

- 89% yield
- t = 0.5 min
- Coiled PFA tubular reactor (volume 479 µL)
- Batch: 3 h reaction time

Scheme 6.26. Stephenson's photoflow functionalization of *N*-aryl tetrahydroisoquinolines.

and the NMR tube, the authors found that the "active volume" of the reactor was directly related to the surface area being irradiated.[43] In seeking a more effective reactor vessel, the Booker-Milburn's design of photomicroflow reactor was applied with 1-W blue LED and 1.6 mm ID. This protocol not only improved the productivity (85% yield), but also improved the turnover frequency (TOF of 30 h^{-1}). With this method at hand, multigram quantity of the aldehyde could be produced, and used for synthesis of several key products such as C-serine glycoamino acid, chloroesters and azidoesters.

The photooxidative generation of the intermediate iminium ions in the C–X reduction by Stephenson also found use in other photoflow processes (Scheme 6.26).[29,44] In the presence of Ru(bpy)$_3$Cl$_2$ and BrCCl$_3$ (to generate CCl$_3$ radical as the H atom abstractor), the iminium ions generated from the SET oxidation of *N*-aryl tetrahydroisoquinolines can serve as good electrophile upon addition by electron-rich heterocycles, cyanide, and alkyne. The typical batch reaction to prepare α-nitroalkyl amines took 3 h. In Stephenson's photoflow reactor, the residence time was 0.5 min.

Besides Ru, Ir and Cu, the Co-complexes have seen their application as photocatalyst in the synthesis of organic molecules. Carreira employed a new strategy to prepare allylic trifluoromethanes using cobalt photocatalyst.[45] Styrene and derivatives were treated with 2,2,2-trifluoroethyl iodide to obtain the corresponding trifluoromethylation products in moderate to high yield (50–80% yield)

	[Co]-catalyst (20 mol%) 2 equiv i-Pr$_2$NEt & I—CF$_3$	
	hv (48 high-power LEDs) CH$_3$CN/DMSO	

Microflow conditions: - Capillary Reactor - Size: 2.2 mm Ø, 8 mL - 0.267 ml/min	Batch:	*t* = 24 hours	76%
	Flow:	***t* = 30 min**	**66%**

Scheme 6.27. Carreira's continuous-flow trifluoroethylation.

in batch. In particular, 66% yield could be achieved for the bulky *tert*-butylstyrene with 20 mol% [Co]-catalysis in a mixture solution of CH$_3$CN/DMSO (1:1). In the continuous microflow reactor with 48 high-power LEDs (equivalent to 37 W, λ_{max} = 465 nm), the reactions with styrene derivatives gave satisfactory productivity, when the flow rate was set at 0.267 mL/min with a 30-min residence time (Scheme 6.27).

Encouraged by these results, the authors revisited alkyl-Heck cyclization, and found that the [Co]-catalyst was also highly efficiently in the intramolecular annulation (Scheme 6.28).[45] With the same catalyst loading (12 mol% [Co]) and same yield (82%), the reaction time was shortened significantly from batch (24 h) to flow (30 min residence time).

6.2.7 *Isomerization and rearrangement*

One ideal synthesis of peptide bond is the native chemical ligation (NCL).[46] One of such methods is the peptide fragment coupling between aldehyde and hydroxylamine.[47] Jamison and coworkers developed a simple and efficient rearrangement of nitrone into amide in continuous-flow mediated by UV light (see Figure 6.8).[48] The core of the microflow reactor consists of a water-cooled quartz immersion well with a 450-W Hg lamp.

With this flow apparatus, nitrones with different aromatic substitution could be converted to corresponding amides in good

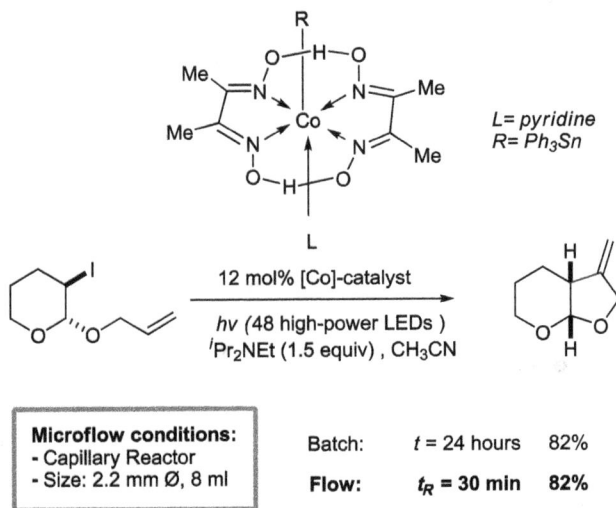

Scheme 6.28. Carreira's continuous-flow alkyl-Heck cyclization.

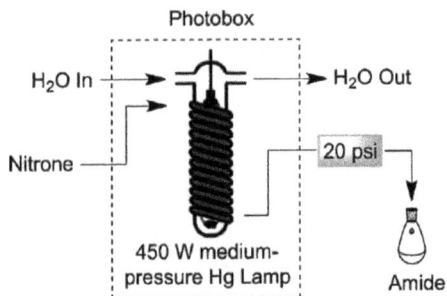

Fig. 6.8. Jamison's photoflow reactor for the rearrangement of nitrones. Copyright © 2013 WILEY-VCH Verlag GmbH & Co. KGaA, Weinheim.

yields (60–90%) at 92°C. The process shortened the reaction time (t_R = 5–20 min) and avoided epimerization of the amide product (Table 6.5).

Encouraged by these results, the authors performed a direct synthesis of amide from aldehyde and hydroxylamine (Scheme 6.29).[48] Nitrone that formed *in situ* from *Ala*-hydroxylamine and *Val*-aldehyde was injected in the flow reactor via syringe pumps, and the desired amide product was obtained in 59% yield with a t_R of

Table 6.5. Scope of the nitrone rearrangement.

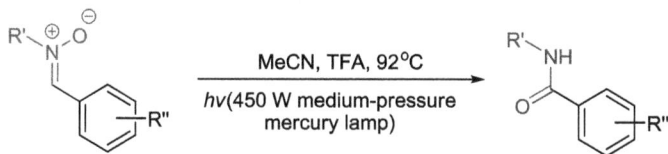

Entry	Substrate [R′ R″]		Time (min)	TFA (equiv)	Yield (%)
1			10	—	83
2	Me	*m*-Me	15	—	70
3			5	0.1	74
4	Me	*o*-Me	15	0.25	60
5	Me	2,4,6-Me	10	0.25	65
6			20	0.25	90
7	Me	*p*-F	15	0.25	84
8			10	0.1	81

(*Continued*)

Table 6.5. (*Continued*)

Entry	Substrate [R' R'']		Time (min)	TFA (equiv)	Yield (%)
9	Be	p-OMe	10	—	87
10			10	0.25	89

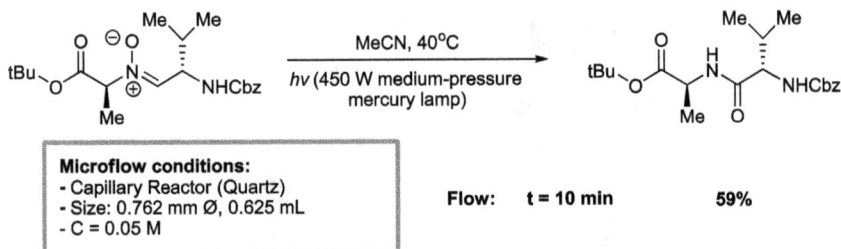

Scheme 6.29. Jamison's flow photochemical rearrangement of nitrone to access dipeptides.

only 10 min. In comparison, the 10 min reaction in batch afforded oxaziridines as the major product (oxaziridines/amide = 7/1).

The Rueping group developed a photocatalytic isomerization of (E)-olefins into (Z)-olefins with Ir catalyst under continuous-flow conditions.[49] The [Ir(ppy)$_2$(bpy)](PF$_6$) acted as a photosensitizers that facilitate the olefin isomerization via an energy-transfer mechanism (Scheme 6.30). In batch, a variety of (Z)-alkenes could be prepared under mild conditions with up to 99:1 Z/E ratio (Table 6.6).

To adopt the catalytic system in flow, the authors first immobilized the Ir catalyst in ionic liquid, and found no activity loss after eight runs in batch. After that, a sustainable flow system was developed, with recycling of the ionic liquid phase where the Ir photocatalyst was immobilized in (Figure 6.9).[49] Almost quantitative conversion of (E)-stilbene to the (Z)-isomer could be achieved with multigram scale.

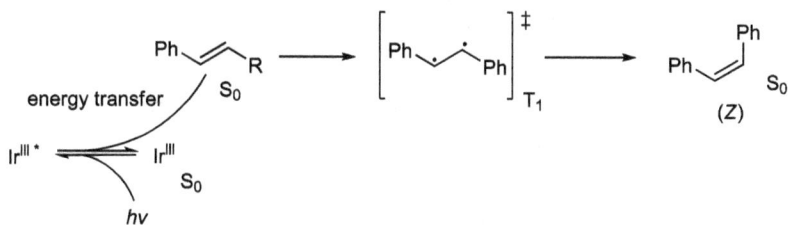

Scheme 6.30. The mechanism of the photoisomerization of olefins.

Table 6.6. Rueping's photoisomerization of olefins.

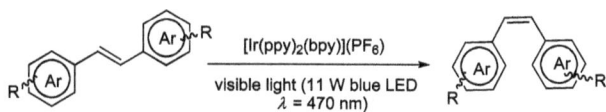

Entry	Products	Z/E	Conv. (%)
1		99:1	98
2		99:1	96
3		99:1	98
4		99:1	98

(*Continued*)

Table 6.6.   (*Continued*)

Entry	Products	Z/E	Conv. (%)
5		9:1	98
6		99:1	98
7		6:1	98
8		99:1	85
9		12:1	93
10		99:1	86

Doxercalciferol is often used to treat chronic kidney disease, but the traditionally insufficient photoisomerization has limited the production of it. Bauta obtained the key intermediate of a vitamin D derivative starting from ergocalciferol in five steps with 20% yield after crystallization.[50] In the key isomerization step, 9-acetylanthracene was utilized as photosensitizer to achieve 50% yield in a photo-flow reactor (Coiled PTFE) irradiated with a 450-W Hg lamp (Scheme 6.31). Non-polar alkane was chosen as solvent and the transformation could be operated at room temperature ($\pm 10°$C). Scale-up related problems could be avoided using this reactor.

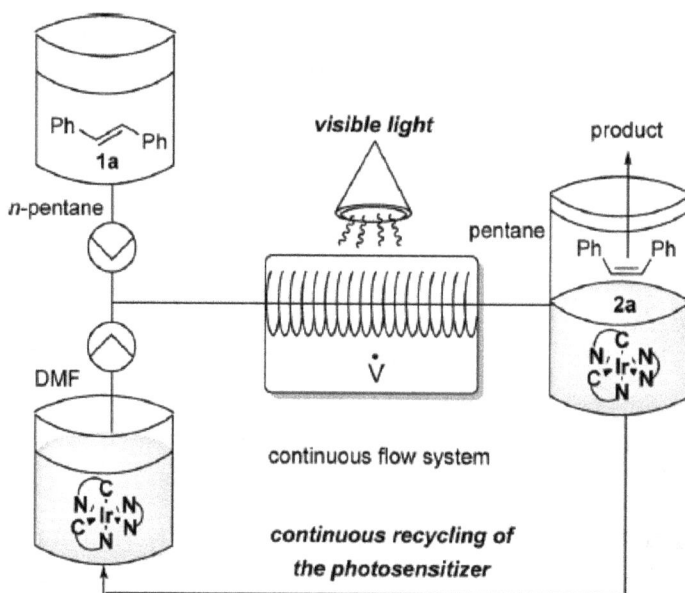

Fig. 6.9. Biphasic photoflow system with recycling of the ionic liquid phase. Copyright © 2015 WILEY-VCH Verlag GmbH & Co. KGaA, Weinheim.

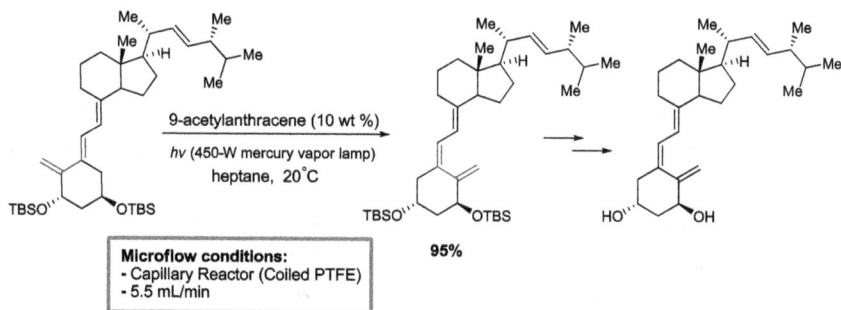

Scheme 6.31. The photoisomerization of vitamin D derivatives.

Booker-Milburn discovered by chance that under UV irradiation, a series of double-bond-containing pyrrole derivatives could undergo intramolecular cyclization and rearrangement, to generate tricyclic aziridine rings.[51] The pyrrole with an electron-withdrawing group at C2 undergoes a $[2 + 2]$ cycloaddition reaction between the $\Delta^{2,3}$ and

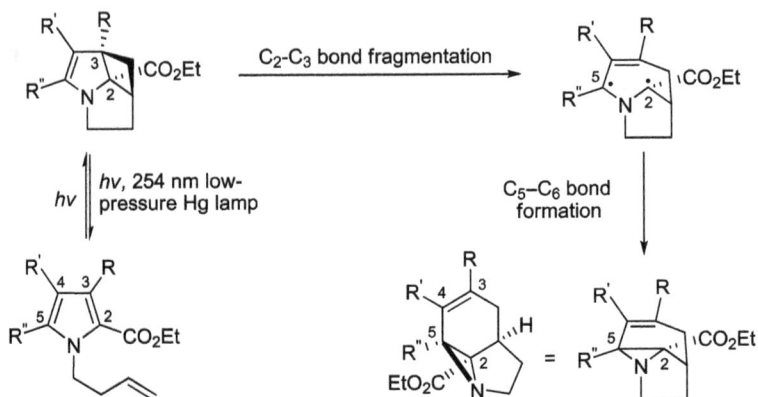

Scheme 6.32. The proposed mechanism of Booker-Milburn's photocycloaddition and rearrangement.

the remote C=C double bond, with a low-pressure Hg lamp. Subsequently, the C2–C3 bond broke and formed diradical. The conjugated recombination of the diradical formed a new bond between C2 and C5, resulting in the aziridine structure (Scheme 6.32).

Astoundingly, with the continuous-flow reactor equipped with a 36-W PL-L lamp, at a flow rate of 9.5 mL/min of the reactant and under irradiation at 254 nm, a yield of 51% could be obtained after 1 h operation. Traditional batch reaction took 6 h to obtain a 59% yield. Similarly, substituting C2 with electron-withdrawing group achieved the same yields of 57% in flow reactor (9.6 mL/min; 1 h) and 56% in batch (6 h) (Scheme 6.33).[51]

Inspired by the pioneering work by Booker-Milburn, Harrowven and his team developed a similar photoflow apparatus.[52] A low-energy 9-W lamp was used to replace the conventional medium-pressure 400–600-W mercury lamp (Figure 6.10). Around the quartz tubing to house the 9-W LED lamp, there were three layers: layer 1 was the PFA double coiled tubing, layer 2 was the aluminum foil to reflect light, and layer 3 was the condenser tubing flowing with cooling water.

Using this flow photochemical reactor, Harrowven's team has achieved good results in the photochemical rearrangement reaction of arylcyclobutenones to 5H-furanone. The reaction

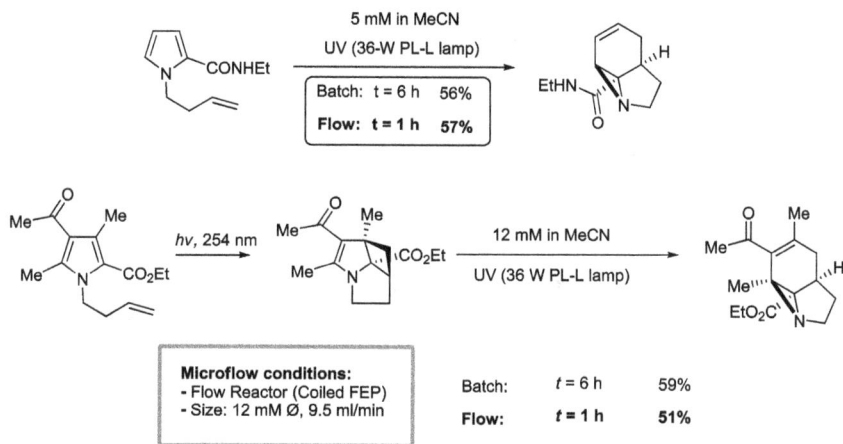

Scheme 6.33. Booker-Milburn's continuous-flow photocycloaddition of pyrrole derivatives.

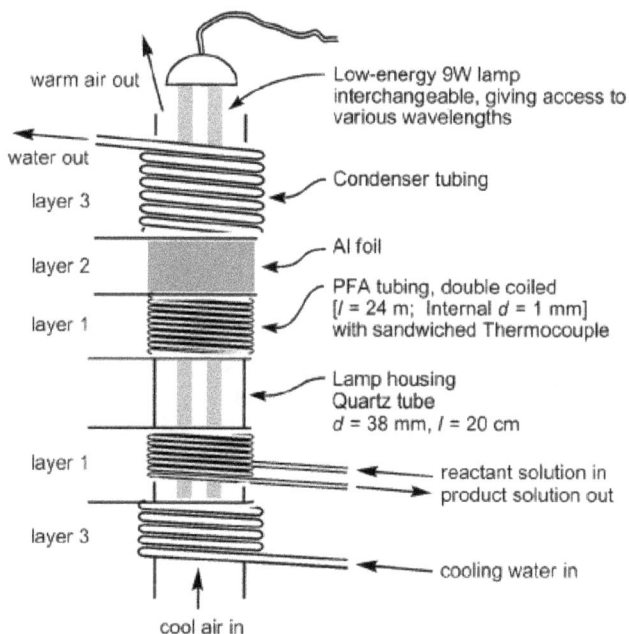

Fig. 6.10. Harrowven's photoflow setup for cycloaddition. Copyright © 2012 WILEY-VCH Verlag GmbH & Co. KGaA, Weinheim.

Table 6.7. Photochemical rearrangement of 4-hydroxycyclobutenones in continuous-flow.

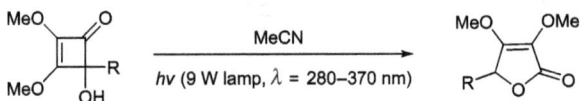

Entry	R	Products	Yield[a,b] (%)
1	Ph		99[a] 27[b]
2	^nBu		95
3	^tBu		96[a] 39[b]
4	Ph-C≡C		95[a] 28[b]
5	o-MeOC$_6$H$_4$		97[b] 51[b]
6	p-Me$_3$SiC$_6$H$_4$		93

(*Continued*)

Table 6.7. (*Continued*)

Entry	R	Products	Yield[a,b] (%)
7	3-Pyridyl		94
8	p-tBuC$_6$H$_4$		92

Note: [a]Photochemical reaction under continuous-flow.
[b]Batch reaction.

with lower-energy 9-W broad-spectrum UVB lamp provided yields of as high as 99% with 90 min residence time. In contrast, only 27% yield was achieved in a 4-h reaction in batch (Table 6.7, Entry 1). Alkyl, pyridine, and aromatic group-substituted analogs were also obtained with satisfactory yields (Entries 2–8, >90% yield).[52]

6.3 Conclusion and Outlook

As a growing technology, the merger of homogeneous photoreactions and continuous-flow chemistry has seen many recent applications. Common advantages of this combination include lower catalyst loading, higher yield, substantial acceleration of reaction rate, and the reduction of side product due to short but effective irradiation time. However, to date, the photoflow reactions have been restricted for lab use. Scaling them up to meet industrial requirement is still in the research stage, since problems such as catalyst recycling need to be addressed. Nevertheless, the development in photoflow chemistry is encouraging, and we foresee that many previously low-yielding batch transformations should regain our attention with the assistance of this new technology.

References

1. N. Hoffmann, *Chem. Rev.* **108**, 1052 (2008).
2. T. Bach and J. P. Hehn, *Angew. Chem. Int. Ed.* **50**, 1000 (2011).
3. N. Hoffmann, *ChemSusChem* **5**, 352 (2012).
4. T. Noël, X. Wang and V. Hessel, *Chim. Oggi* **31**(3), 10 (2013).
5. Y.-H. Su, N. J. W. Straathof, V. Hessel and T. Noël, *Chem. Eur. J.* **20**, 10562 (2014).
6. R. L. Hartman, *Org. Process Res. Dev.* **16**, 870 (2012).
7. R. Porta, M. Benaglia and A. Puglisi, *Org. Process Res. Dev.* **20**, 2 (2016).
8. K. Gilmore and P. H. Seeberger, *Chem. Rev.* **14**, 410 (2014).
9. C. Birr, W. Lochinger, G. Stahnke and P. Lang, *Liebigs Ann. Chem.* **763**, 162 (1972).
10. B. D. A. Hook, W. Dohle, P. R. Hirst, M. Pickworth, M. B. Berry and K. I. Booker-Milburn, *J. Org. Chem.* **70**, 7558 (2005).
11. D. Šterk, M. Jukič and Z. Časar, *Org. Process Res. Dev.* **17**, 145 (2013).
12. D. Cantillo, O. de Frutos, J. A. Rincon, C. Mateos and C. O. Kappe, *J. Org. Chem.* **79**, 223 (2014).
13. D. Cantillo, O. de Frutos, J. A. Rincon, C. Mateos and C. O. Kappe, *J. Org. Chem.* **79**, 8486 (2014).
14. J. D. Nguyen, E. M. D'Amato, J. M. R. Narayanam and C. R. J. Stephenson, *Nat. Chem.* **4**, 854 (2012).
15. J. D. Nguyen, B. Reiss, C. Dai and C. R. J. Stephenson, *Chem. Commun.* **49**, 4352 (2013).
16. F. R. Bou-Hamdan and P. H. Seeberger, *Chem. Sci.* **3**, 1612 (2012).
17. E. Sugiono and M. Rueping, *Beilstein J. Org. Chem.* **9**, 2457 (2013).
18. B. Shen, M. W. Bedore, A. Sniady and T. F. Jamison, *Chem. Commun.* **48**, 7444 (2012).
19. B. Liégault, D. Lee, M. P. Huetis, D. R. Stuart and K. Fagnou, *J. Org. Chem.* **73**, 5022 (2008).
20. Q. Lefebvre, M. Jentsch and M. Rueping, *Beilstein J. Org. Chem.* **9**, 1883 (2013).
21. A. C. Hernandez-Perez, A. Vlassova and S. K. Collins, *Org. Lett.* **14**, 2988 (2012).
22. A. C. Hernandez-Perez and S. K. Collins, *Angew. Chem. Int. Ed.* **52**, 12696 (2013).
23. A. C. Hernandez-Perez, A. Caron and S. K. Collins, *Chem. Eur. J.* **21**, 16673 (2015).
24. M. D. Lainchbury, M. I. Medley, P. M. Taylor, P. Hirst, W. Dohle and K. I. Booker-Milburn, *J. Org. Chem.* **73**, 6497 (2008).
25. T. Fukuyama, Y. Kajihara, Y. Hino and I. Ryu, *J. Flow Chem.* **1**, 40 (2011).
26. K. Terao, Y. Nishiyama, H. Tanimoto, T. Morimoto, M. Oelgemoeller and K. Kakiuchi, *J. Flow Chem.* **2**, 73 (2012).
27. K. Asano, Y. Uesugi and J.-I. Yoshida, *Org. Lett.* **15**, 2398 (2013).
28. A. C. Gutierrez and T. F. Jamison, *Org. Lett.* **13**, 6414 (2011).

29. J. W. Tucker, Y. Zhang, T. F. Jamison and C. R. J. Stephenson, *Angew. Chem. Int. Ed.* **51**, 4144 (2012).
30. S. L. Poe, M. A. Cummings, M. P. Haaf and T. D. McQuade, *Angew. Chem. Int. Ed.* **45**, 1544 (2006).
31. T. Horie, M. Sumino, T. Tanaka, Y. Matsushita, T. Ichimura and J.-I. Yoshida, *Org. Process Res. Dev.* **14**, 405 (2010).
32. A. Yavorskyy, O. Shvydkiv, N. Hoffmann, K. Nolan and M. Oelgemoeller, *Org. Lett.* **14**, 4342 (2012).
33. X. Wang, G. D. Cuny and T. Noël, *Angew. Chem. Int. Ed.* **52**, 7860 (2013).
34. E. Bremus-Köbberling, A. Gillner, F. Avemaria, C. Rethore and S. Bräse, *Beilstein J. Org. Chem.* **8**, 1213 (2012).
35. F. R. Bou-Hamdan, F. Levesque, A. G. O'Brien and P. H. Seeberger, *Beilstein J. Org. Chem.* **7**, 1124 (2011).
36. D. Cantillo, O. de Frutos, J. A. Rincon, C. Mateos and C. O. Kappe, *Org. Lett.* **16**, 896 (2014).
37. M. Neumann and K. Zeitler, *Org. Lett.* **14**, 2658 (2012).
38. J. W. Tucker, J. M. R. Narayanam, S. W. Krabbe and C. R. J. Stephenson, *Org. Lett.* **12**, 368 (2010).
39. M. D. Konieczynska, C. Dai and C. R. J. Stephenson, *Org. Biomol. Chem.* **10**, 4509 (2012).
40. F. W. Schmidtmann, T. E. Benedum and G. J. McGarvey, *Tetrahedron Lett.* **46**, 4677 (2005).
41. R. W. Franck and M. Tsuji, *Acc. Chem. Res.* **39**, 692 (2006).
42. A. Nuzzi, A. Massi and A. Dondoni, *Org. Lett.* **10**, 4485 (2008).
43. R. S. Andrews, J. J. Becker and M. R. Gagné, *Angew. Chem. Int. Ed.* **51**, 4140 (2012).
44. D. B. Freeman, L. Furst, A. G. Condie and C. R. J. Stephenson, *Org. Lett.* **14**, 94 (2012).
45. L. M. Kreis, S. Krautwald, N. Pfeiffer, R. E. Martin and E. M. Carreira, *Org. Lett.* **15**, 1634 (2013).
46. P. E. Dawson, T. W. Muir, I. Clarklewis and S. B. H. Kent, *Science* **266**, 776 (1994).
47. S. I. Medina, J. Wu and J. W. Bode, *Org. Biomol. Chem.* **8**, 3405 (2010).
48. Y. Zhang, M. L. Blackman, A. B. Leduc and T. F. Jamison, *Angew. Chem. Int. Ed.* **52**, 4251 (2013).
49. D. C. Fabry, M. A. Ronge and M. Rueping, *Chem. Eur. J.* **21**, 5350 (2015).
50. B. G. Anderson, W. E. Bauta and W. R. Cantrell, *Org. Process Res. Dev.* **16**, 967 (2012).
51. K. G. Maskill, J. P. Knowles, L. D. Elliott, R. W. Alder and K. I. Booker-Milburn, *Angew. Chem. Int. Ed.* **52**, 1499 (2013).
52. D. C. Harrowven, M. Mohamed, T. P. Goncalves, R. J. Whitby, D. Bolien and H. F. Sneddon, *Angew. Chem. Int. Ed.* **51**, 4405 (2012).

Chapter 7

Heterogeneous Photoreactions in Continuous Flow

Timothy Noël

Department of Chemical Engineering & Chemistry
Laboratory for Micro Flow Chemistry & Process Technology
Eindhoven University of Technology
De Rondom 70, Eindhoven 5612AP, The Netherlands
t.noel@tue.nl

7.1 Introduction

Multiphase reactions deal with more than one phase (i.e. gas, liquid or solid). This typically includes gas–liquid reactions and liquid–liquid reactions but also reactions utilizing heterogeneous catalysts. Common examples in the chemical industry are oxidations, hydrogenations, halogenations and phase-transfer catalysis. It is often observed that in such cases the reaction rate is determined by mass transfer limitations from one phase to the other. These limitations can be overcome in microreactors as microdevices provide a large and well-defined interfacial area (up to 9000 $m^2\,m^{-3}$).[1]

7.2 Gas–Liquid Photochemical Reactions

Several microreactor designs can be distinguished to enable gas–liquid photochemical reactions: (i) capillary microreactors; (ii) falling film microreactors; and (iii) membrane microreactors (see Figure 7.1).[2,3] In a capillary microreactor, gas–liquid flow usually results in a Taylor flow regime (also called slug or segmented flow). This flow regime is characterized by gas bubbles and liquid slugs. In these segments, toroidal vortices are established which result in an improved mixing and thus fast mass transfer between the liquid and the gas phase. When the velocity of the gas phase increases, an annular flow regime will be encountered (typically the gas velocity should be 10 times the liquid phase velocity) where the gas phase flows in the center of the tube, while the liquid phase forms a thin film at the channel walls. Both flow regimes display an excellent contact between the gas and the liquid phase. In addition to the optimal irradiation of the reaction mixture, drastically reduced reaction times are typically observed in microreactors for gas–liquid photochemical transformations.[4,5] The second design is the falling film microreactor in which a thin film is established due to gravitational forces. The gas phase can be subsequently directed counter- or co-currently in front of the liquid phase. The last design is the membrane microreactor where the two phases are flowing in two channels which are separated by a gas permeable membrane.[6]

7.2.1 *Oxidation reactions*

The use of oxygen in combination with photosensitizers and light irradiation results in the generation of singlet oxygen.[7,8] Singlet oxygen is a more reactive form of oxygen which provides many opportunities in organic synthesis.[5] However, its use in batch is often plagued by overoxidation processes. In contrast, the reaction time can be controlled efficiently in flow resulting in less byproduct formation. The generation of singlet oxygen is one of the most investigated reactions in continuous-flow gas–liquid photochemistry.

An interesting example is the use of singlet oxygen to enable [4 + 2] cycloadditions. This synthetic strategy was used to prepare ascaridole in flow (Scheme 7.1).[9] The photoreactor consisted of a glass chip (50 × 150 μm channels, 375 μL volume) and was irradiated by a 20-W tungsten lamp. More than 80% conversion was observed in <5 s residence time.

Artemisinin is one of the most important antimalarials on the market. The key step of the synthetic process is a photochemical singlet oxygen reaction. A flow protocol was developed

Single channel microreactor:

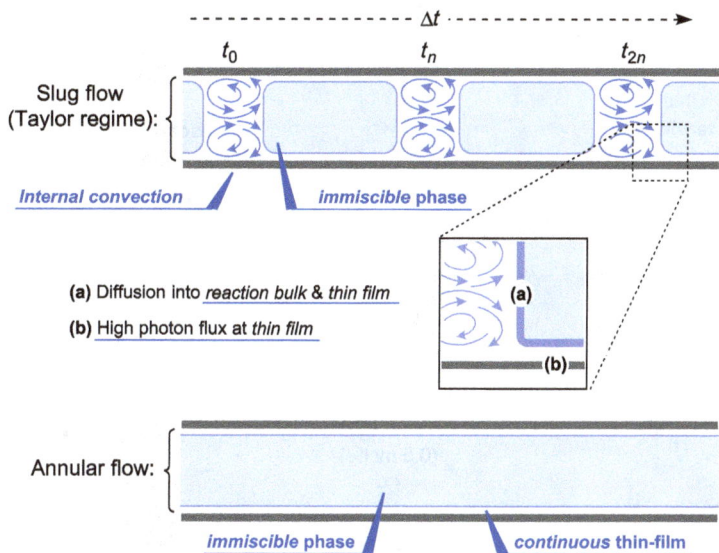

Falling Film Microreactor:

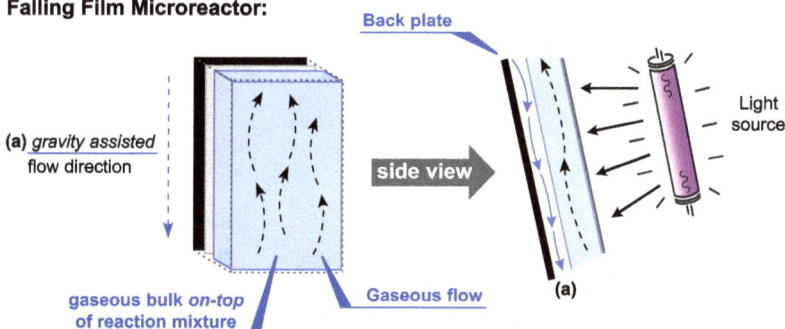

Membrane Microreactor

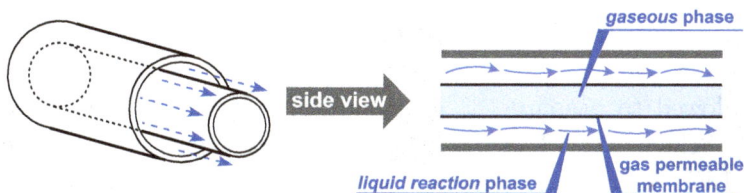

Fig. 7.1. Overview of the different gas–liquid photomicroreactors: single channel capillary microreactor, falling film microreactor and membrane microreactor.

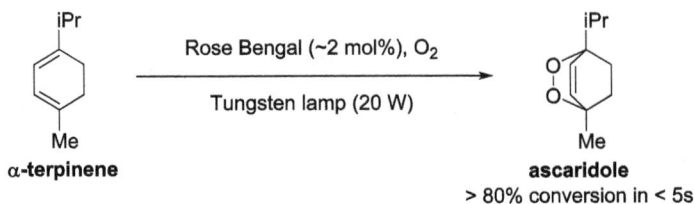

Scheme 7.1. Synthesis of ascaridole via a [4 + 2] photooxygenation of α-terpinene in flow.

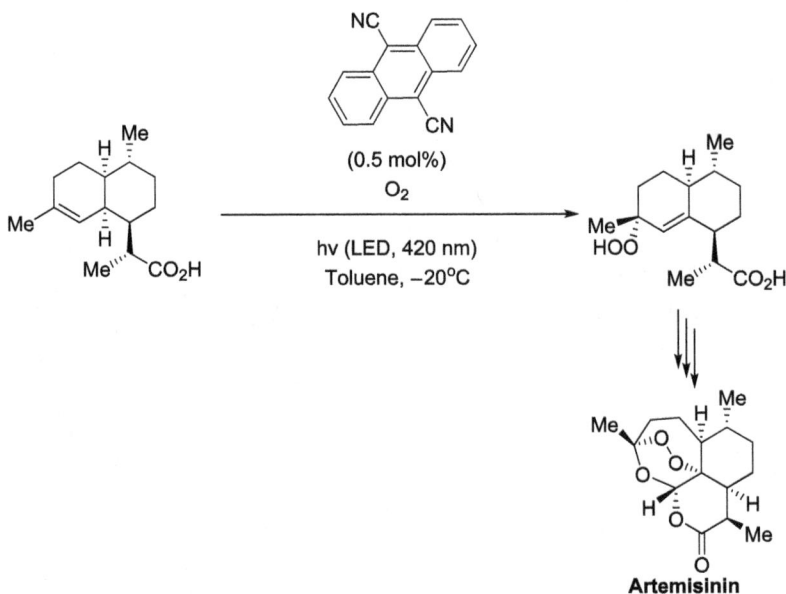

Scheme 7.2. Singlet oxygen generation as key photochemical oxidation step for the synthesis of artemisinin in continuous flow.

by Seeberger *et al.* to prepare artemisinin starting from dihydroartemisinic acid.[10,11] The entire flow process required 11.5 min and allowed to prepare 165 g artemisinin per day (65–70% overall yield) (Scheme 7.2). The photochemical step was carried out in a capillary FEP reactor (750 µm inner diameter, 7.5 mL volume) which was subjected to LED irradiation (12 W, 420 nm). Dicyanoanthracene was used as the photosensitizer to generate singlet oxygen.

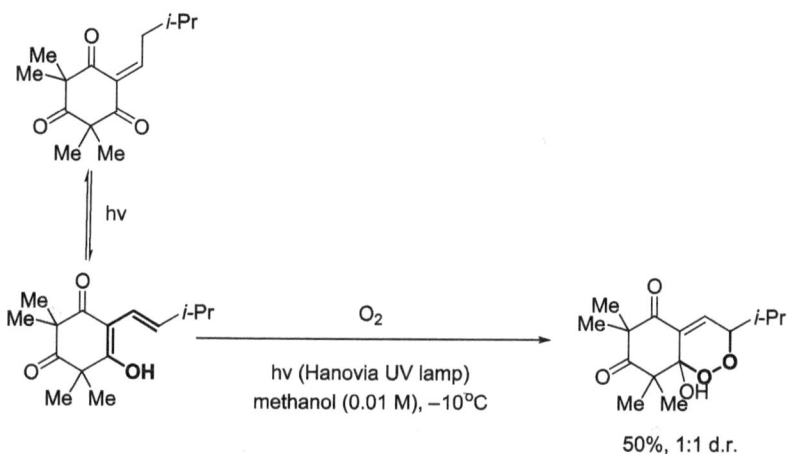

Scheme 7.3. Continuous-flow [4+2] cycloaddition with singlet oxygen en route to rhodomyrtosone A.

In the synthesis of rhodomyrtosone A, a photochemical [4 + 2] cycloaddition with singlet oxygen in flow was used by Porco *et al.* to prepare sufficient amounts of the peroxide intermediate (Scheme 7.3).[12] The starting material was dissolved in methanol, saturated with oxygen and introduced in a PFA capillary microreactor (1.5 mm internal diameter, 3.1 mL). The capillary was wrapped around a UV lamp and cooled to −10°C. The corresponding endoperoxide was obtained as a 1:1 mixture of diastereomers in 50% yield.

Kappe *et al.* have reported the use of singlet oxygen in flow to convert 5-hydroxymethylfurfural into highly functionalized building blocks (Scheme 7.4).[13] The reactor consists of PFA capillary (1.0 mm internal diameter, 10 mL volume) and is irradiated by a 60-W compact fluorescent light (CFL) bulb. The substrates and Rose Bengal as photocatalyst were dissolved in *i*-PrOH/H$_2$O (1:1) and merged with an oxygen stream. At 17 bar, full conversion was observed in 40 min residence time.

Seeberger *et al.* developed a photochemical oxidative cyanation of primary and secondary amines (Scheme 7.5).[14] The photoreactor consists of an FEP capillary (750 μm internal diameter, 7.5 mL volume) which was irradiated with 420 nm LEDs. In this transformation,

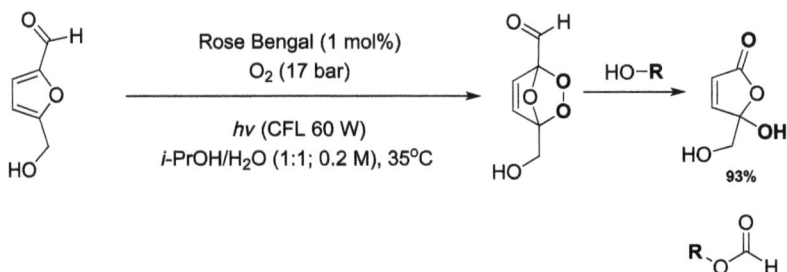

Scheme 7.4. Continuous-flow [4 + 2] cycloaddition of 5-hydroxymethylfurfural with singlet oxygen.

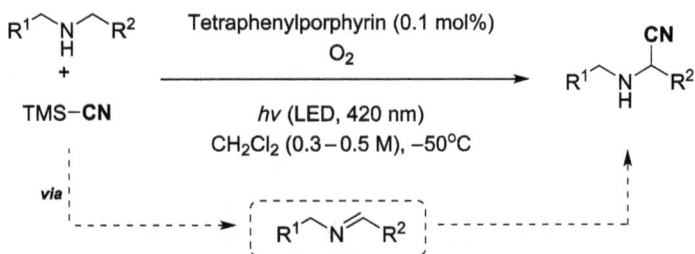

Scheme 7.5. Continuous-flow oxidative cyanation of primary amines via photocatalytic singlet-oxygen.

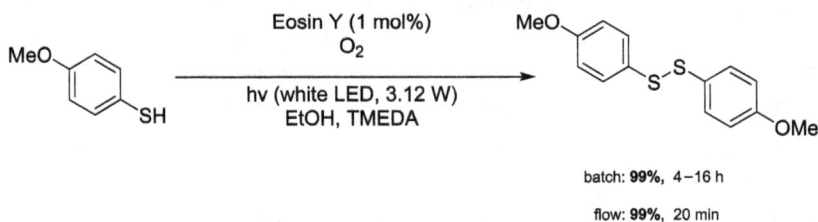

Scheme 7.6. Continuous-flow aerobic oxidation of thiols to disulfides.

singlet oxygen is used to generate imines which are subsequently trapped with trimethylsilyl cyanide to produce α-aminonitriles.

An aerobic oxidation of thiols to disulfides was developed by Noël *et al.* Eosin Y was found to be the optimal photocatalyst (Scheme 7.6).[15] TMEDA was added as a base and ethanol was used as a green solvent. The photomicroreactor consisted of PFA capillary

(750 μm internal diameter, 950 μL volume) which was irradiated with white LEDs.[16] A variety of different thiols were converted in flow to the corresponding disulfide in less than 20 min residence time in high purity. Moreover, the method could be extended to unprotected peptides; an intramolecular disulfide formation resulted in the active form of the non-apeptidic oxytocin in only 200 s irradiation time. This reaction was subsequently used as a model reaction to study fundamental engineering processes. It was found that mass transfer limitations could be completely overcome in a Taylor flow regime.[17] Furthermore, the photonic efficiency of the photomicroreactor was 0.66. Such high values can be explained by the matching dimensions between the LED light source and the capillary.[18]

7.2.2 *Trifluoromethylation reactions*

The trifluoromethyl group (CF_3) is an important structural motive to alter the properties of pharmaceuticals, agrochemicals and materials. More specifically, incorporation of CF_3 in pharmaceuticals results in an increased lipophilicity, bioavailability and metabolic stability. Consequently, in recent years, numerous methods have been developed to allow trifluoromethylation of organic molecules. However, many tailor-made trifluoromethylating reagents are quite expensive, narrow in scope and/or difficult to prepare. One exception is CF_3I which can be engaged readily in radical processes.[19] The major drawback of this reagent is, however, the gaseous nature of this reagent.

Noël *et al.* have developed continuous-flow methods to allow efficient and accurate handling of CF_3I.[16] Mass flow controllers dose the gaseous reactant into the liquid stream and allow to control precisely the stoichiometry. Due to the high solubility of this gas in many solvents (e.g. 1.6 M in acetonitrile), a homogeneous reaction stream is observed in flow.[20]

The trifluoromethylation of five-membered heterocycles was achieved in a PFA capillary microreactor (750 μm internal diameter, 480 μL volume) (Scheme 7.7).[21] $Ru(bpy)_3Cl_2$ was the optimal photocatalyst and was irradiated with blue LEDs (5.88 W) in flow. In the presence of TMEDA (N, N, N', N'-tetramethyl-1,2-diaminoethylene)

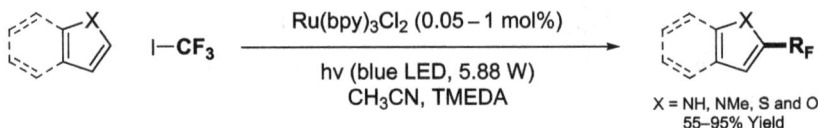

Scheme 7.7. Continuous-flow trifluoromethylation of five-membered heterocycles with CF_3I.

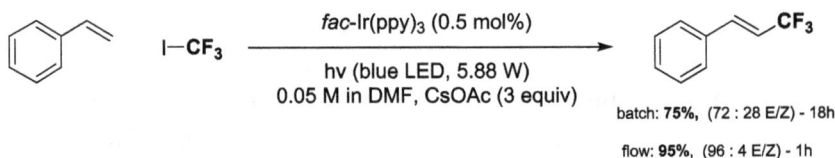

Scheme 7.8. Continuous-flow trifluoromethylation of styrenes with CF_3I.

as an electron donor, a reductive quenching pathway is followed.[22] The quantum yield of the reaction is $\Phi = 0.55 \pm 0.02$ indicating that the photocatalytic trifluoromethylation of heterocycles proceeds via a non-chain pathway. All substrates could be fully converted in flow in good to excellent yield (55–95% yield) in short reaction times (8–16 min versus 12–72 h in batch). Eosin Y was also demonstrated to enable this reaction but with a reduced efficiency compared to $Ru(bpy)_3Cl_2$.[23]

A similar setup was used to allow for trifluoromethylation of styrenes (Scheme 7.8).[24] The reaction could be accelerated from 18 h in batch to 60 min in flow. Interestingly, the E/Z selectivity was greatly improved from 72:28 in batch to 96:4 in flow. This selectivity increase can be attributed to the reduced reaction times observed in flow, which allows to avoid photoinduced E/Z interconversion.

Next, Noël *et al.* have developed a continuous-flow method for the trifluoromethylation of thiols to establish S–CF_3 bonds.[25] A broad scope of different thiols could be efficiently trifluoromethylated in batch and flow. Interestingly, the amount of CF_3I gas could be reduced from 4 equiv in batch to 1.1 equiv in flow. In batch, the CF_3I gas escapes from the liquid solution into the head space and diffuses only marginally back into the solution. Recently, the same

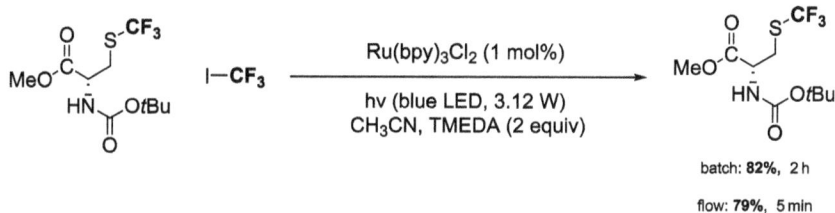

Scheme 7.9. Continuous-flow trifluoromethylation of cysteines with CF_3I.

authors also extended the scope to the trifluoromethylation of cysteine and cysteine-containing peptides (Scheme 7.9).[26] Interestingly, for this transformation, a high quantum yield of $\Phi = 126$ was found, clearly indicating the presence of a chain-propagating single electron transfer.

7.3 Heterogeneous Photocatalysis

The use of heterogeneous photocatalysts is interesting as it allows to recuperate and recycle the catalysts efficiently. However, it brings also some additional engineering complications for the design of a photomicroreactor. Heterogeneous photocatalysts can be used as a suspension, immobilized on the reactor walls or fixated in a packed bed. Furthermore, complex phenomena occur at the catalyst surface itself, including diffusion phenomena, adsorption, desorption and reaction (Figure 7.2). As a result of the high surface-to-volume ratio of microreactors, heterogeneous photocatalysts can be efficiently excited resulting in high reaction rates. Furthermore, intensified mass transfer characteristics, encountered in such devices, can also overcome most transport limitations.

7.3.1 *Titanium dioxide*

Titanium dioxide (TiO_2) is the most widely used semiconductor photocatalyst. However, its use is mostly reserved for applications in the water treatment industry.[27] TiO_2 absorbs mainly in the UV region but the absorbance can be shifted to the visible light region via doping or surface complexes.

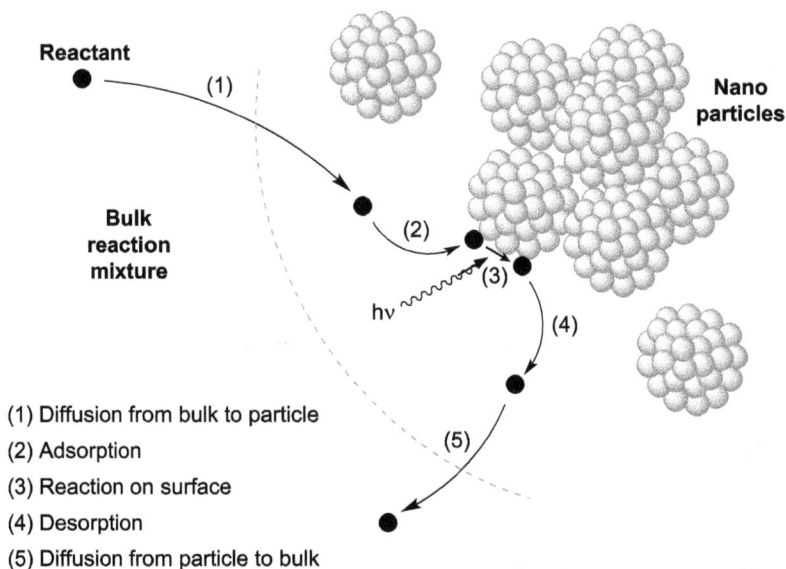

Fig. 7.2. Transport phenomena involved in heterogeneous photocatalysis.

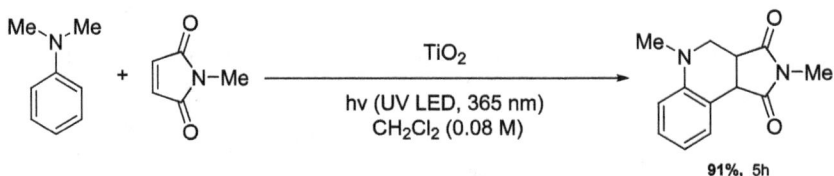

Scheme 7.10. TiO_2 photocatalytic addition–cyclization reaction of N-methylmaleimide and N, N-dimethylaniline in flow.

Kappe *et al.* have developed a continuous-flow protocol for the tandem addition–cyclization of N-methylmaleimide and N, N-dimethylaniline using suspended colloidal TiO_2 nanocrystals (Scheme 7.10).[28] The reaction was carried out in a PFA capillary microreactor (750 μm internal diameter, 650 μL) and irradiated with UV LEDs (365 nm). The target compound could be obtained after recirculating the reaction mixture for 5 h in the microreactor.

To overcome the issues of handling solid laden flows (e.g. clogging), TiO_2 can be immobilized on the channel walls, i.e. a

Batch: t = 5 hours (TiO$_2$/Pt)   85%   3%

Flow: t = 90 sec   98%   0%

(a)

Batch: t = 60 min   14% (47% ee)

Flow: t = 52 sec   14% (50% ee)

(b)

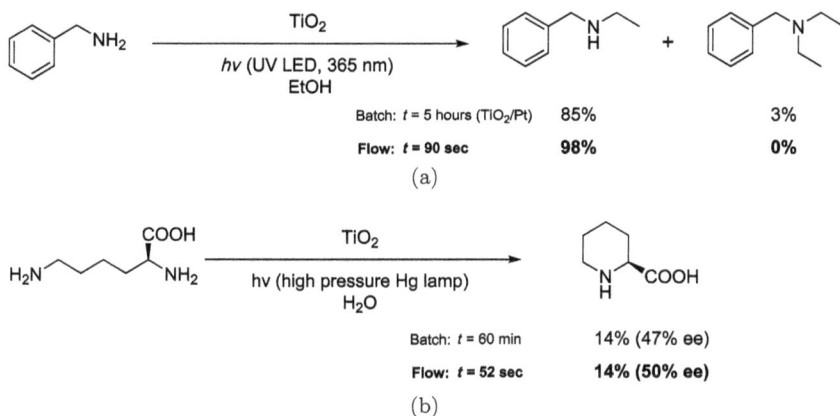

Scheme 7.11. Photocatalytic transformations in TiO$_2$ wall-coated photomicroreactors. (a) Photocatalytic N-ethylation of benzylamine in a quartz microreactor with a TiO$_2$ layer excited with 365 nm UV LEDs. (b) Synthesis of *L*-pipecolinic acid in a Pyrex microreactor with a TiO$_2$ layer excited with a high pressure mercury lamp.

wall-coated photomicroreactor. Due to high surface-to-volume ratios, large interfacial areas can be obtained in microreactors. Examples of such reactors in synthesis are the *N*-ethylation of benzylamine (Scheme 7.11(a))[29] and the synthesis of *L*-pipecolinic acid (Scheme 7.11(b)).[30] In both cases, substantial accelerations of the reaction kinetics have been observed.

Noël *et al.* have developed a batch and flow protocol for the photocatalytic aerobic oxidation of thiols to disulfides using TiO$_2$ and visible light (Scheme 7.12).[31] TiO$_2$ was loaded in a packed-bed reactor and the segmented oxygen–liquid stream was directed over it. The addition of TMEDA was crucial to cause aggregation of the TiO$_2$ nanoparticles and thus to prevent clogging and leaching of the catalyst. The reaction could be accelerated in flow (5 min versus 8 h in batch) and no catalyst degradation was observed for >28 h.

7.3.2 *Mesoporous graphitic carbon nitride*

Blechert *et al.* have used mesoporous graphitic carbon nitride (mpg-C$_3$N$_4$) to enable radical cyclizations in flow (Scheme 7.13).[32] The heterogeneous photocatalyst was loaded in a packed-bed reactor and

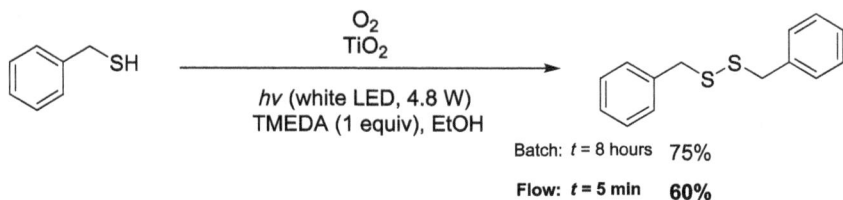

Scheme 7.12. Photocatalytic aerobic oxidation of thiols to disulfides using TiO$_2$ in a continuous-flow packed-bed reactor and visible light.

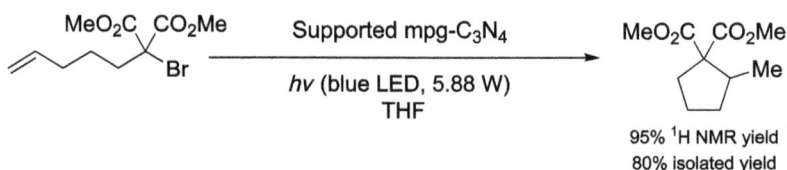

Scheme 7.13. Radical cyclizations using mpg-C$_3$N$_4$ in a continuous-flow packed-bed reactor.

tetrahydrofuran was the optimal solvent. A wide variety of substrates could be efficiently converted in flow. Interestingly, the catalyst remained stable for a long time (60–70 reaction cycles).

7.4 Conclusion

As shown in this chapter, continuous-flow microreactors are ideal reactors to facilitate heterogeneous photochemical reactions. This is due to the excellent interfacial contact, which can overcome mass transfer limitations, and the optimal irradiation of the reaction medium encountered in such reactors. Another advantage is the ease of scale-up of such complicated reaction conditions. Furthermore, heterogeneous catalysts can be immobilized in a flow reactor providing opportunities for facile catalyst recuperation/recycling and to increase Turn Over Numbers (TONs). To date, only small set of different multiphase photoreactions have been performed in such reactors. Consequently, it is fair to claim that many more examples will appear in the future. It is remarkable that only a limited amount of heterogeneous catalysts have been applied in organic synthetic photochemistry. Given the importance of such catalysts in other fields, we believe that this will rapidly change. In this regard, another interesting opportunity is to develop a continuous-flow protocol for the fast screening of heterogeneous photocatalysts in flow.

References

1. T. Noël, Y. Su and V. Hessel, *Top Organomet. Chem.* **57**, 1 (2016).
2. T. Noël and V. Hessel, *ChemSusChem* **6**, 405 (2013).
3. C. J. Mallia and I. R. Baxendale, *Org. Process Res. Dev.* **20**, 327 (2016).
4. Y. Su, N. J. W. Straathof, V. Hessel and T. Noël, *Chem. Eur. J.* **20**, 10562 (2014).
5. D. Cambie, C. Bottecchia, N. J. W. Straathof, V. Hessel and T. Noël, *Chem. Rev.* **116**, 10276 (2016).
6. M. Brzozowski, M. O'Brien, S. V. Ley and A. Polyzos, *Acc. Chem. Res.* **48**, 349 (2015).
7. M. DeRosa, *Coord. Chem. Rev.* **233–234**, 351 (2002).
8. P. R. Ogilby, *Chem. Soc. Rev.* **39**, 3181 (2010).
9. R. C. R. Wootton, R. Fortt and A. J. de Mello, *Org. Process Res. Dev.* **6**, 187 (2002).
10. F. Levesque and P. H. Seeberger, *Angew. Chem. Int. Ed.* **51**, 1706 (2012).
11. D. Kopetzki, F. Levesque and P. H. Seeberger, *Chem. Eur. J.* **19**, 5450 (2013).
12. A. Gervais, K. E. Lazarski and J. A. Porco, *J. Org. Chem.* **80**, 9584 (2015).
13. T. S. A. Heugebaert, C. V. Stevens and C. O. Kappe, *ChemSusChem* **8**, 1648 (2015).
14. D. B. Ushakov, K. Gilmore, D. Kopetzki, D. T. McQuade and P. H. Seeberger, *Angew. Chem. Int. Ed.* **53**, 557 (2014).
15. A. Talla, B. Driessen, N. J. W. Straathof, L.-G. Milroy, L. Brunsveld, V. Hessel and T. Noël, *Adv. Synth. Catal.* **357**, 2180 (2015).
16. N. J. W. Straathof, Y. Su, V. Hessel and T. Noël, *Nat. Protoc.* **11**, 10 (2016).
17. Y. Su, V. Hessel and T. Noël, *AIChE J.* **61**, 2215 (2015).
18. Y. Su, A. Talla, V. Hessel and T. Noël, *Chem. Eng. Technol.* **38**, 1733 (2015).
19. S. Barata-Vallejo, S. M. Bonesi and A. Postigo, *Org. Biomol. Chem.* **13**, 11153 (2015).
20. A. Studer, *Angew. Chem. Int. Ed.* **51**, 8950 (2012).
21. Y. Su, K. P. L. Kuijpers, N. Koenig, M. Shang, V. Hessel and T. Noël, *Chem. Eur. J.* **22**, 12295 (2016).
22. N. J. W. Straathof, H. P. L. Gemoets, X. Wang, J. C. Schouten, V. Hessel and T. Noël, *ChemSusChem* **7**, 1612 (2014).
23. N. J. W. Straathof, D. J. G. P. van Osch, A. Schouten, X. Wang, J. C. Schouten, V. Hessel and T. Noël, *J. Flow Chem.* **4**, 12 (2014).
24. N. J. W. Straathof, S. E. Cramer, V. Hessel and T. Noël, *Angew. Chem. Int. Ed.* 2016, DOI: 10.1002/anie.201608297R2.
25. N. J. W. Straathof, B. J. P. Tegelbeckers, V. Hessel, X. Wang and T. Noël, *Chem. Sci.* **5**, 4768 (2014).

26. C. Bottecchia, X.-J. Wei, K. P. L. Kuijpers, V. Hessel and T. Noël, *J. Org. Chem.* **81**, 7301 (2016).
27. S. Malato, P. Fernandez-Ibanez, M. I. Maldonado, J. Blanco and W. Gernjak, *Catal. Today* **147**, 1 (2009).
28. M. Baghbanzadeh, T. N. Glasnov and C. O. Kappe, *J. Flow Chem.* **3**, 109 (2013).
29. Y. Matsushita, N. Ohba, S. Kumada, K. Sakeda, T. Suzuki and T. Ichimura, *Chem. Eng. J.* **135**, S303 (2008).
30. G. Takei, T. Kitamori and H.-B. Kim, *Catal. Commun.* **6**, 357 (2005).
31. C. Bottecchia, N. Erdmann, P. M. A. Tijssen, L.-G. Milroy, L. Brunsveld, V. Hessel and T. Noël, *ChemSusChem* **9**, 1781 (2016).
32. M. Woznica, N. Chaoui, S. Taabache and S. Blechert, *Chem. Eur. J.* **20**, 14624 (2014).

Chapter 8

Photoflow Material Synthesis

Benjamin Wenn* and Thomas Junkers[*,†,‡]

*Polymer Reaction Design Group
Institute for Materials Research (IMO)
Universiteit Hasselt
Martelarenlaan 42, 3500 Hasselt, Belgium

†IMEC Associated Lab IMOMEC
Wetenschapspark 1, 3590 Diepenbeek, Belgium
‡thomas.junkers@uhasselt.be

8.1 Introduction

Continuous-flow processing is the method of choice in the large-scale production of everyday materials. Many bulk (co)polymers are produced on ton scale in large flow reactors. Yet, in the field of precision materials engineering, no significant interest was until recently found for continuous production. Especially in the polymer field, very significant advances have been made in the past 20 years and a virtually endless number of high-precision materials with different composition, topology, microstructure, functionality and dispersity can nowadays be accessed (see Figure 8.1).

Techniques like reversible addition fragmentation chain transfer polymerization (RAFT),[1,2] atom transfer radical polymerization (ATRP),[3,4] or nitroxide-mediated polymerization (NMP),[5,6] in conjunction with the advent of modular macromolecular design strategies as made available from the *Click* chemistry concept[7] have led to

Macromolecular Structure

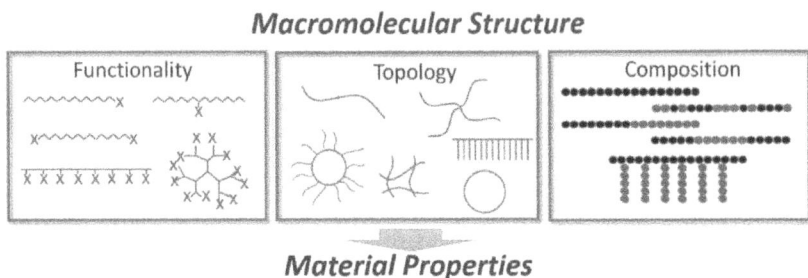

Material Properties

Fig. 8.1. Structure-property relationship in polymeric materials.

a true synthetic revolution in material science. The precise synthesis of (polymeric) materials allows for detailed studies into structure–property relationships and in consequence for a rational design of novel materials based on *ab initio* considerations. While highly powerful from a synthetic point of view, however, upscaling remains a significant problem for the aforementioned techniques. Scaling up of the polymerizations (or polymer modifications) often leads to a loss of structural integrity, broadening of macromolecular weight distributions and increased numbers of defects. In many aspects, flow chemistry can offer suitable solutions to the upscaling problem. Especially the increasing use of micro and mesoscale reactors allow not only for simply upscale, but even for improved reaction conditions and outcomes compared to classical batch processes.

Polymerizations are inherently not simple to carry out in continuous processes. With polymerization, a viscosity increase is unavoidably connected, which leads to potential blockages, increased reactor fouling, changed flow profiles and significant pressure drops. Yet, by careful design of reactors — using dilute conditions or by working in droplet phases — these problems can be overcome. Due to the excellent heat transfer often associated with micro and mesoreactors, stable temperature conditions can be achieved with ease in flow. The exothermicity of polymerizations is often a crucial aspect, even if it is often conveniently ignored by synthetic bench chemists. Well known by engineers, efficient heat dissipation or removal is required to keep stable reaction conditions. On the relatively small synthetic scale of bench chemists, heat dissipation is often assumed to be unproblematic and no precaution is taken to prevent it, though a significant heat-up even of small polymerization containers can be often observed. For example, the single-electron transfer polymerization (SET-LRP),[8,9] a technique usually referred to as a "room temperature polymerization", usually heats up to 50–60°C even in the absence of any external heat source when being carried out on a few mL scale. Recently, Derboven *et al.*[10] showed how the quality of the obtained polymer product is significantly increased in microstructured flow reactors simply due to the fact that side reactions are under better control due to isothermal processing. In this case, a thermal RAFT polymerization was studied by experiment and by comparing results with high-level Monte Carlo simulations. A match between simulation, flow and batch experimental results could only be achieved when the observed temperature increase of 30 K that occurred in the batch reactions due to the reaction exotherm was taken into account.

The distinct advantage of increased product quality has been exploited for a number of (homogenous) polymerization reactions and polymer modifications. RAFT, ATRP, NMP and SET-LRP polymerizations[8,10−30] have all been performed in continuous flow numerous times, each time showing that the product quality regarding dispersity, endgroup fidelity and overall yield in time was improved. Increasing reactor diameters can lower the beneficial effect, however, products never fall behind the associated batch process. Also classical living polymerizations have been studied, and flow reactors were identified to be ideal tools to carry out anionic and cationic polymerizations.[18,31,32] Similar advantages as for the controlled radical polymerization techniques were observed. As for polymer modifications, advantages were also identified here when processing those in flow (even though this can be largely attributed to the inherently easier optimization flow reactors.[33] and not to an isothermicity effect). For example thiol-ene modifications of RAFT polymers were optimized to reaction times of several minutes (compared to hours in batch).[29] Also block copolymer formation via copper-catalyzed azide–alkyne click reactions could be carried out in less than 1 h, while usually being done in batch as overnight reactions.[34] That flow polymerizations are superior to their batch counterparts is not only evident when comparing dispersity data, but is also directly observed by the fact that flow reactions aid significantly in the sequential design of multiblock copolymers. To this end, pentablock copolymers were obtained by consecutive RAFT flow polymerizations, even though the according batch process did not allow for any block extension after formation of a triblock copolymer due to the larger loss of active endgroups.[28] By coupling of flow reactors, multiblock copolymers can also be directly obtained in one-step processes — again in a fashion that is not reachable in batch.[18,35]

In summary, continuous-flow production of high-precision polymer materials is highly advantageous. Yet, a significantly larger advantage is even observed when switching flow reactors from thermal to photochemical mode.[36] Photoreactions — as will be described below in detail — benefit in a way flow processing that easily surpasses the above described effects.[37] Flow photoreactions allow for enormous increases in reaction rates and are able to eliminate side-product formation significantly. This effect is not routed to isothermicity, but is due to the excellent light illumination that is reached when internal flow diameters are kept low.

When optical path lengths are short, full illumination of the reaction mixture in its full volume can be reached. In batch reactors, light intensity gradients are unavoidable (see Figure 8.2). Following Beer–Lambert's law, light is absorbed by the UV active ingredients in the mixture, quickly lowering the intensity for the following volume increment. Typically, UV light does not penetrate much deeper than few centimeters into a reaction mixture. Thus, while batch processing required rigorous stirring in order to make sure that all reaction components reach the vessel surface (and hence the illuminated area) within a certain amount of time, practically full illumination and hence full excitation of the active ingredients is reached in the flow process. The consequence is largely reduced reaction times for the photoreactions. At the same time, the overall light intensity can

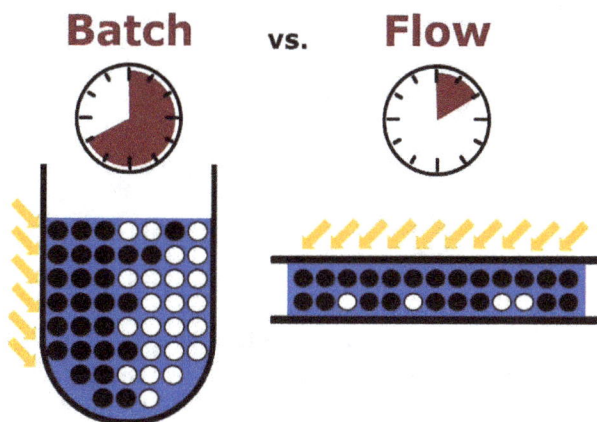

Fig. 8.2. Schematic representation of the "flow advantage" of photoreactions. While batch reactors exhibit light intensity gradients due to Beer–Lambert's law, practically full excitation of the whole reaction volume is reached in flow.

often be reduced significantly, which avoids photodegradation reactions that are otherwise often observed.[22,38−42]

In this chapter, we will focus on UV-induced reactions for precision material synthesis carried out in continuous micro and mesosized flow reactors. The main emphasis will be given to polymer reactions, but also formation of inorganic or hybrid nanoparticles will be discussed, where similar efficiency gains are seen as sketched above.

8.2 Continuous Photoinitiated Controlled Radical Polymerization

8.2.1 *Transition metal mediated polymerizations*

A photoinduced copper-mediated controlled radical polymerization in a continuous-flow reactor was performed by Junkers and co-workers. They polymerized linear methacrylate homo and diblock copolymers via copper-mediated polymerization in a micro and milliflow reactor.[15] Later, they showed the possibility to polymerize methyl methacrylate and methacrylate/acrylate block copolymers with the same technique in continuous-photoflow reactors (Scheme 8.1).[25]

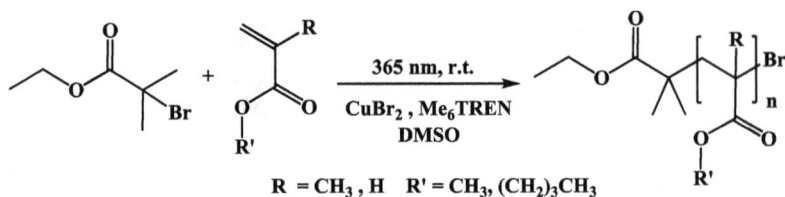

Scheme 8.1. Reaction scheme for (meth)acrylate polymerization via photoinitiated copper-mediated polymerization.

For the acrylate polymerization, they used a micro as well as a milliflow reactor to show the easy scalability of the photoinitiated polymerization. As microflow reactor, a Chemtrix BV Labtrix® Start R2.2 system with a glass chip microreactor was employed. The reaction solutions were delivered via a syringe pump from two gas-tight glass syringes. As light source, a 100-W high-pressure mercury vapor short arc lamp was employed. The larger milliflow reactor consisted of tightly wrapped PFA tubing around a quartz cooling mantel for a 400-W medium-pressure UV lamp. An HPLC pump was used to pump the reaction mixture into the reactor. The reaction mixtures were degassed by purging with nitrogen before delivering into the reactor for both setups.

In a first step, they synthesized poly(methyl acrylate) with three different targeted molar masses via photoinitiated copper-mediated polymerization in the milliflow reactor. For all three experiments, a linear increase in molecular weight with increasing monomer conversion was observed (Figure 8.3). High molar masses with low dispersities and conversion was reached in reaction times of maximal 20 min. This is significantly shorter than previously reported batch photopolymerizations of the same kind.[43] For all three targeted molecular masses, a high level of control was observed. The kinetic first-order plot shows linear behavior and indicate a constant radical concentration (Figure 8.4). A limitation in the flow reaction was that only intermediate molecular masses could be targeted since an increase in molecular weight leads to a concomitant viscosity increase, which may cause reactor blockages. Nevertheless, the employed simple reactor already gave rise to a production up to

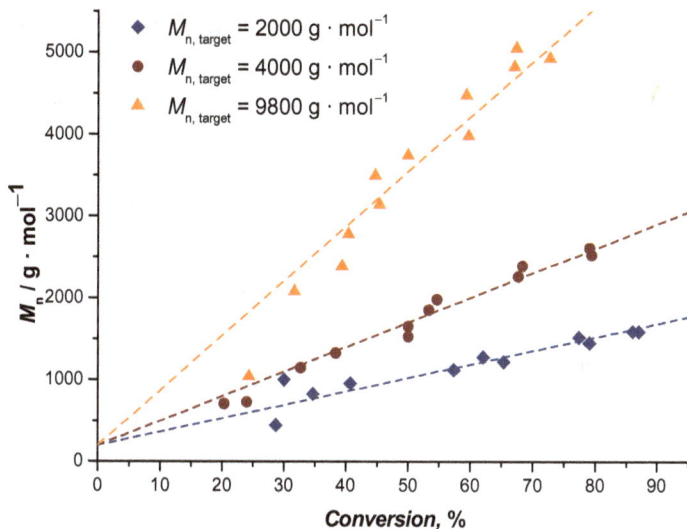

Fig. 8.3. Development of the number average molecular weight of poly(methyl acrylate) synthesized via photoinitiated copper-mediated polymerization in a continuous milliflow reactor. (Reprinted with permission from Ref. 15. Copyright © 2014 The Royal Society of Chemistry.)

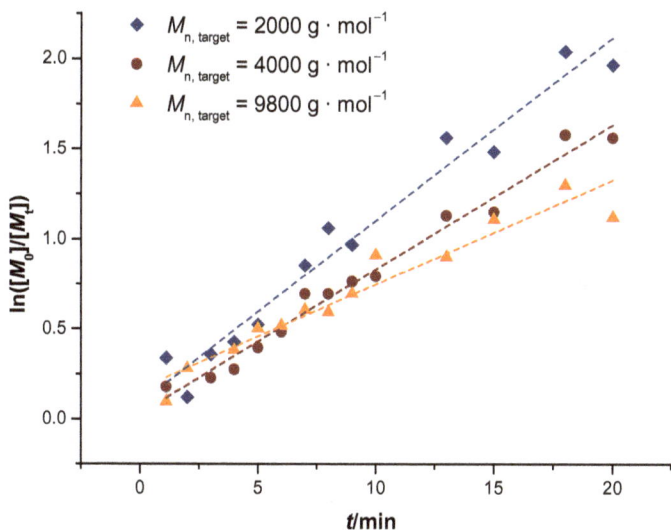

Fig. 8.4. Kinetic first-order plots for the synthesized poly(methyl acrylate) in a continuous-photoflow reactor via copper-mediated polymerization. (Reprinted with permission from Ref. 15. Copyright © 2014 The Royal Society of Chemistry.)

60 g of polymer a day, which is quite significant when comparing the commercial value of such well-controlled polymers.

To demonstrate the simple scalability of the photopolymerization the same reaction was performed in a microflow reactor. In this reactor, the identical photoinitiated copper-mediated polymerization of methyl acrylate in DMSO was performed. A molecular weight of $4000 \, g \, mol^{-1}$ was targeted as described above. Monomer conversion of about 80% were achieved in 20 min reaction time, while likewise a good control over the reaction was retained. A comparison of the first-order kinetic plots for the milli and microflow reactions display good linearity with a slight difference in the slope (Figure 8.5), which can be explained by the differences in the light source used. The y-axis offset is explained by preliminary polymerization occurring already in the feed syringes, which was, however, inconsequential for the product obtained.

Junkers and coworkers also polymerized poly(methyl methacrylate) in the identical milliflow reactor via photoinitiated copper-mediated polymerization.[25] Also here a significant reduction

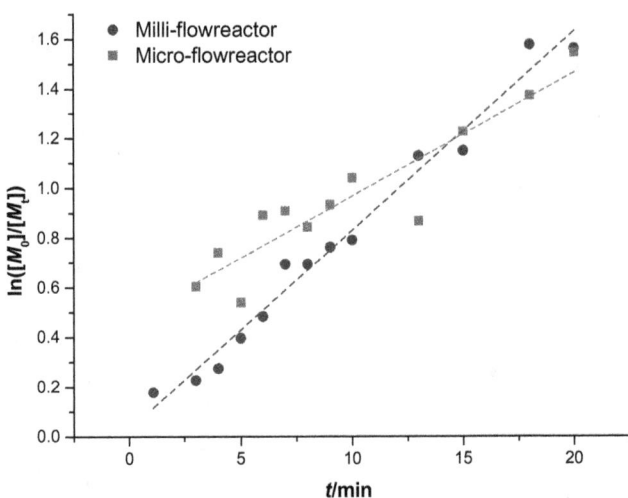

Fig. 8.5. Comparison of the first-order kinetic plots of methyl acralyte polymerization in milli and microflow reactors via photoinitiated copper-mediated polymerization. (Reprinted with permission from Ref. 15. Copyright © 2014 The Royal Society of Chemistry.)

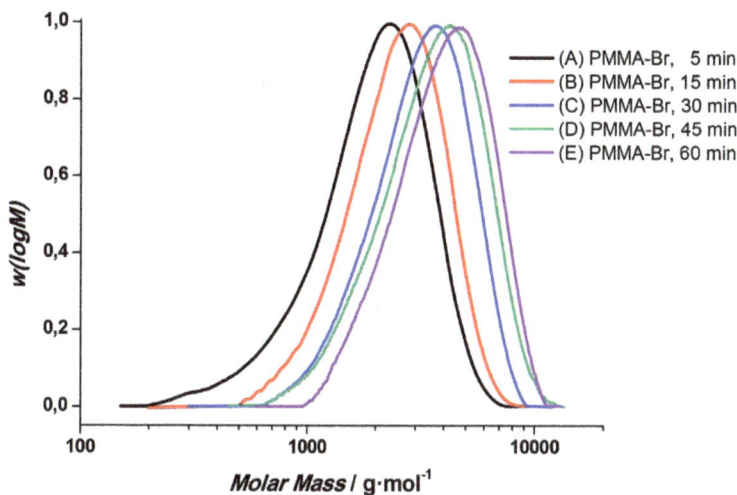

Fig. 8.6. SEC traces of poly(methyl methacrylate) at different reaction times. The polymerization was performed in a continuous-milliflow reactor with the photoinitiated copper-mediated technique. (Reprinted with permission from Ref. 25. Copyright © 2015 The Royal Society of Chemistry.)

of reaction time for the batch to flow comparison was observed. After workup, poly(methyl methacrylate) with high molar masses and low dispersities was obtained (Figure 8.6).

Further, also the ability to form block copolymers based on acrylate/methacrylate monomers was demonstrated. For the acrylate polymerization, poly(methyl acrylate) was used after purification as macroinitiator and was subsequently chain extended with *n*-butyl acrylate in the microflow reactor system. The increase in molecular weight during chain extension at different reaction times is depicted in Figure 8.7(a). Similar results are observed for the chain extension of poly(methyl methacrylate) with methyl acrylate in the milliflow system (Figure 8.7(b)). In both cases, low dispersities are maintained, underpinning the high control that is exerted over the reaction.

Next to linear homo and block copolymers, also more complex polymer architectures can be made via photoflow copper-mediated polymerization. The same approach can be applied to obtain star-shaped homo and multiblock copolymers. They synthesized their star-shaped polymers via the core-first approach.[44,45] with up to 21

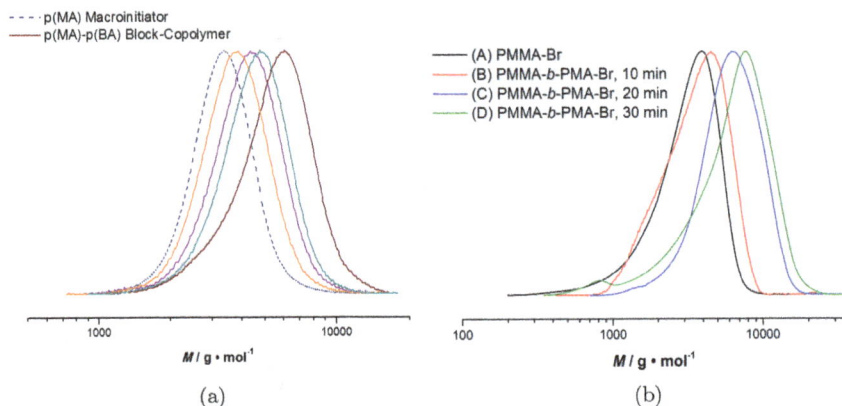

Fig. 8.7. Evolution of molecular weight in block copolymer synthesis in continuous-photoflow reactors via photoinitiated copper-mediated polymerization. (a) Poly(methyl acrylate)-*b*-poly(*n*-butyl acrylate). (Reprinted with permission from Ref. 15. Copyright © 2014 The Royal Society of Chemistry.) (b) Poly(methyl methacrylate)-*b*-poly(methyl acrylate). (Reprinted with permission from Ref. 25. Copyright © 2015 The Royal Society of Chemistry.)

arms per molecule. Up to eight blocks were attached to each arm. Each block had masses between $500\,\mathrm{g\,mol^{-1}}$ and $5100\,\mathrm{g\,mol^{-1}}$ per arm. Star-shaped polymers with masses above $100{,}000\,\mathrm{g\,mol^{-1}}$ were synthesized.[46]

Melker *et al.* performed recently a similar reaction in photoflow. They polymerized methyl methacrylate catalyzed by Ir(ppy)$_3$ as inorganic photoredox system.[17] In their study, they tested the influence of different tubing materials on the polymerization, namely tubing made out of perfluoro alkoxyalkane (PFA), fluorinated ethylene propylene (FEP), Tefzel® and Halar®. These materials differ most significantly in their gas permeability, which has a direct influence on the polymerization rate as oxygen inhibits radical polymerizations. Unsurprisingly, the reaction rates are increasing with a decreasing oxygen permeability (Table 8.1). If the oxygen permeability decreases by a factor of 35, the reaction rate, however, only increases by 30%. The obtained dispersities stay in a comparable range. Thus, gas permeability has critical impact on air-sensitive reactions, its role must, however, also not be overestimated.

Table 8.1. Comparison of the polymerization rate for photoinitiated methyl methacrylate polymerization in different tubing. The reaction was catalyzed by $Ir(ppy)_2$ with a reaction time of 220 min.

Tubing	Oxygen permeability[a]	Conv. (%)	M_n (g mol^{-1})	$Đ$
PFA	881	31	3300	1.23
FEP	748	37	3500	1.15
Tefzel®	100	41	5300	1.24
Halar®	25	42	5400	1.21

Note: [a]Oxygen permeability is reported in units of cm^3/100 in^2 24 h atm mil^{-1} at 25°C (Upchurch Scientific).[17]

Another example for the positive effect of photoflow reactors on yield quality in polymerizations was reported from Detrembleur and colleagues.[23] In their work, poly(vinyl acetate) and poly(vinyl acetate-*co*-1-octene) copolymers were made via photoinitiated cobalt-mediated polymerizations. Polymerizations were performed in a Labtrix Start 2.2® system equipped with a 19.5-μL reactor chip, the same as was used for the copper-mediated polymerizations. For reaction times above 20 min a FlowSyn glass mixer block from Uniqsis with an internal volume of 2 mL was utilized. Syringe pumps with 1 or 10 mL gas-tight glass syringes were used to pump the degassed reaction solutions into the reactors. Their study focused on vinyl acetate polymerization in continuous-flow reactors under thermal activation with and without additional photoactivation. Kinetic analysis of experiments in ethyl acetate in a batch reactor, thermal flow reactor and thermal and photoflow reactor underlines clearly the advantages of using photoflow reactors. While a thermal-activated polymerization has comparable reaction rates to the batch process, the photoflow process at 40°C shows a dramatic increase in reaction rate, even if in the flow process the monomer had to be diluted with ethyl acetate. Under equal reaction conditions, a batch reaction shows by far the slowest polymerization rate within the tested conditions (Figure 8.8).

After proving the positive effect of utilizing photoflow reactors on the reaction rate, the influence on product quality was analyzed. Cobalt-mediated radical polymerizations are known to be difficult

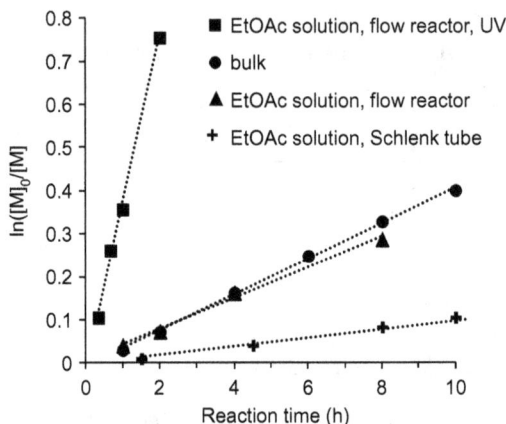

Fig. 8.8. Kinetic first-order plots with time for vinyl acetate polymerization with a cobalt-based catalyst. (Reprinted with permission from Ref. 23. Copyright © 2015 The Royal Society of Chemistry.)

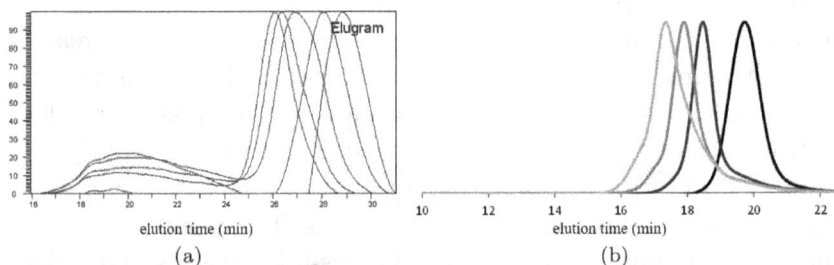

Fig. 8.9. SEC traces of polyvinyl acetate made via cobalt-mediated polymerization in a photobatch (a) (reprinted with permission from Ref. 47. Copyright © 2012 The Royal Society of Chemistry) and photoflow process (b) (reprinted with permission from Ref. 23. Copyright © 2015 The Royal Society of Chemistry).

to handle as batch reactions. The carbon–cobalt bond is photo and thermally labile. Especially for monomers forming very reactive radicals, such as vinyl acetate, handling in UV-batch process is difficult and leads to significant side reactions. The more constant and stable reaction conditions in the continuous-flow reactors had an important impact also on the quality of the obtained polymer. A comparison of the SEC data of poly(vinyl acetate) synthesized in batch and in flow directly demonstrated that side reactions could be avoided in flow (Figure 8.9). No high molecular weight polymers resulting from unwanted branching or other side reactions are present in the SEC chromatograms of the photoflow product (Figure 8.9(b)). Thus, the cobalt-mediated photopolymerization is a nice example for material synthesis where utilization of flow reactors not only led to an improved time-yield correlation, but also to the complete avoidance of side products.

In additional experiments, Detrembleur and coworkers also used a photomicroflow reactor to determine the copolymerization parameters for vinyl acetate with the less reactive 1-octene.[23] Normally, vinyl acetate/1-octene copolymerizations are slow and synthesis of well-defined copolymers with a high olefin content remains limited.[48] The reactions were performed at 40°C under UV irradiation with a reaction time of 8 min to limit monomer conversion and to avoid monomer composition drifts. Experiments with different vinyl acetate/1-octene compositions were completed and the results fitted with the non-linear Mayo–Lewis equation. Reactivity ratios of $r_{\text{VAc}} = 1.73$ and $r_{\text{1-Oct}} = 0.01$ were obtained underlining the general higher incorporation of vinyl acetate than 1-octene.

8.2.2 Reversible addition-fragmentation chain transfer polymerization

The employment of metal complexes in continuous-flow processes can create difficulties due to their ability to precipitate, which can lead to severe blockages in the reactor and which should be avoided as much as possible. One option for metal-free controlled radical polymerization is to use trithiocarbonates (TTCs) as polymerization mediators in the RAFT process (Scheme 8.2). In thermal RAFT

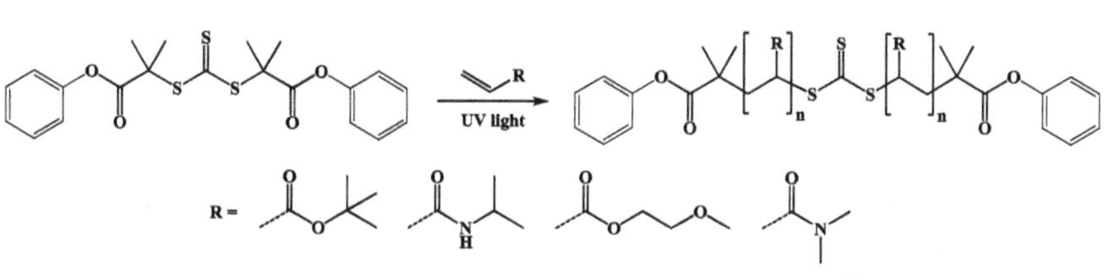

Scheme 8.2. Photo RAFT reaction employing a trithiocarbonate control agent and different monomers.

polymerizations, a conventional initiator is required to keep the RAFT process ongoing. Under UV light, the TTC can absorb light and form radicals by scission of the carbon–sulfur bond, and in this way support polymerization.

Chen and Johnson performed a photoRAFT polymerization in a flow reactor consisting out of commercially available components.[24] Since photoRAFT reactions are sensitive to oxygen, materials with a minimized gas permeability were used. Two reactors were build out of Halar® tubing which was wrapped once around a UV lamp with a peak emission of 352 nm and once around a glass bottle which was placed in the middle of identical lamps. To inject the degassed reaction solution, syringe pumps with stainless steel syringes were used. The reactors were equipped with a back pressure regulator to prevent backflow before the sample was collected (Figure 8.10).

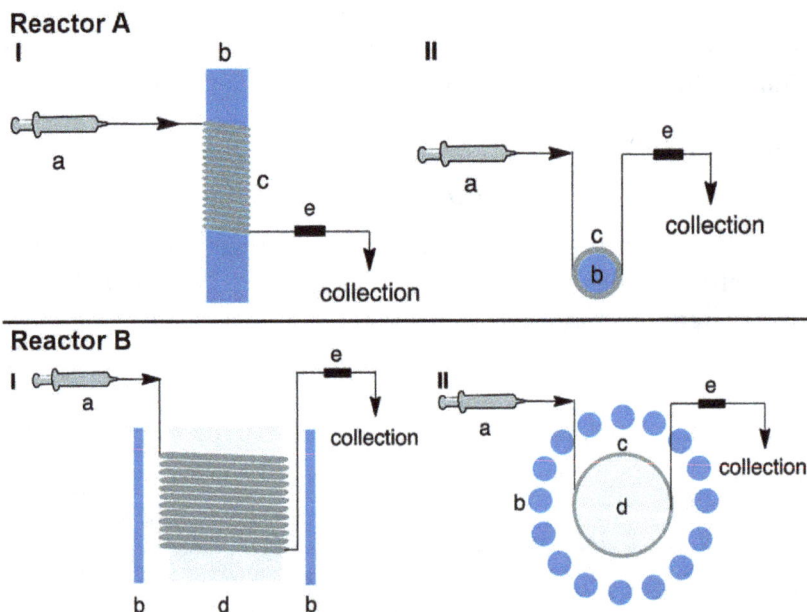

Fig. 8.10. Continuous-flow reactor setup used by Chen and Johnson for photoRAFT polymerization. (I) View from the side and (II) view from the top. (a) Stainless steel syringe, (b) 352 nm peak emission UV lamp, (c) Halar® reactor tubing, (d) glass bottle and (e) back pressure regulator. (Reprinted with permission from Ref. 24. Copyright © 2015 The Royal Society of Chemistry.)

As a first polymerization, a mixture of NiPAAM and TTC with a ratio of 250:1 was made and dissolved in acetonitrile. The mixture was deoxygenated with three freeze-pump-thaw cycles and transferred to the syringes in nitrogen atmosphere. Reactions performed in reactor A showed a linear relationship between the obtained number average molecular weight (M_n) and monomer conversion as expected for a well-controlled RAFT process. Also low dispersities were obtained, underpinning the good control over the reaction. A rate 4-fold reaction rate increase of the flow reaction compared to the identical batch reaction was observed; in flow a reaction time of 60 min is required to reach 80% conversion with the given setup. In the identical batch reaction around 240 min are needed to reach a monomer conversion of 74%.

For higher monomer to TTC ratios (500:1) they observed at reaction times above 30 min a steep increase in dispersity (>1.4). A reason for this loss of control is the consumption of TTC in the photoinitiation. To avoid such decomposition, they created reactor setup B (Figure 8.10) by wrapping the tubing around a glass bottle and placing it in the middle of 8 UV lamps. The greater distance between lamps and reactor leads to a reduced light intensity. Due to the reduced light intensity, the monomer conversion also is reduced slightly from 65% to 50% for a reaction time of 30 min, however, also the unwanted side reactions are lowered. For reactor B, also a linear dependency between M_n and the monomer conversion is observed and lower dispersities were measured (<1.2).

In a next step, both reactors were used to synthesize polymers with molecular weights up to 100,000 g mol^{-1} with different monomers (Table 8.2), nicely demonstrating that photoflow reactors give access to polymer materials with significant molecular weight.

As last step, a triblock copolymer was synthesized in reactor B. For the chain extension a macro-TTC initiator synthesized in reactor A was used. After a reaction time of 60 min in reactor B, a triblock copolymer was obtained (as a symmetrical TTC was employed triblock copoylmers are directly obtained from the homopolymer) with an increased molecular mass was obtained (Figure 8.11). This proofs

Table 8.2. Overview of polymers synthesized via photoflow RAFT polymerizations.[24]

	Monomer	Reactor	Time (min)	Conversion (%)	M_n (g mol^{-1})	$Ð$
1	*t*BA	A	60	81	29,800	1.21
2	*t*BA	B	90	82	54,200	1.19
3	EGMEA	A	60	88	30,300	1.21
4	EGMEA	B	80	81	55,800	1.17
5	DMA	A	40	87	24,900	1.11
6	DMA	B	80	84	44,300	1.18
7	DMA	B	40	55	105,800	1.22

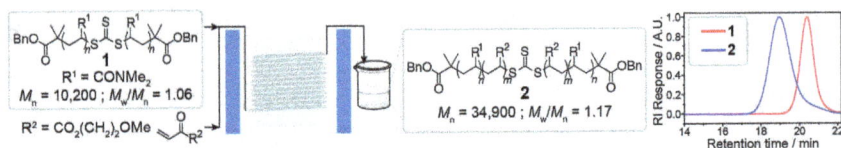

Fig. 8.11. Synthesis of a pDMA-b-pEGMEA-b-pDMA triblock copolymer in a photoflow reactor using RAFT. (Reprinted with permission from Ref. 24. Copyright © 2015 The Royal Society of Chemistry.)

that the photoRAFT polymerization directly allows to reactivate the macroRAFT agents and to perform efficient chain extensions.

8.3 Polymer Modification in Photoflow Reactors

A further application of photoflow reactors is the modification of polymeric material. Conradi and Junkers demonstrated a significant increase in reaction rate by employing a photoflow reactor compared to batch for an endgroup photomodification.[21] As photoflow reactor, a PFA tubing with 0.75 mm inner diameter was wrapped around a quartz-glass cooling mantle of a 400-W medium-pressure UV lamp.

The reaction that was tested for endgroup modification was a [2 + 2] ene–enone cycloaddition. Therefore, a polymer containing an enone endgroup (maleimide) was reacted with a series of alkenes. The polymer bearing the maleimide endgroup (poly(butyl acrylate)) was synthesized stepwise via ATRP. An *N*-hydroxysuccinimide-functionalized initiator was employed in the ATRP, which could then

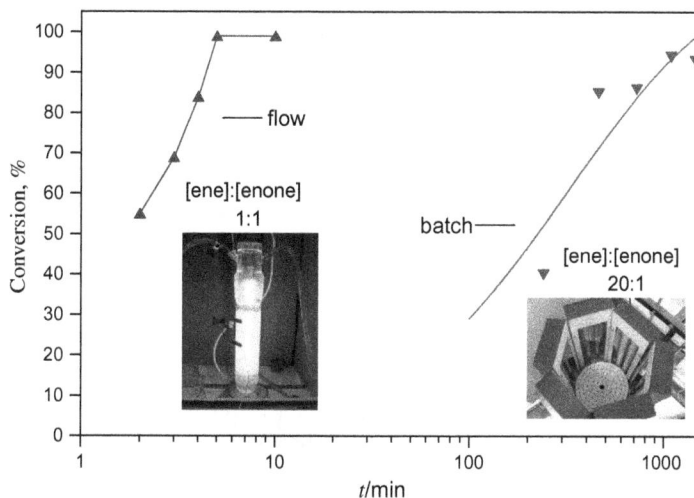

Fig. 8.12. Reaction rate comparison for a [2 + 2] alkene–enone cycloaddition carried out in a batch and flow reactor. In flow, equimolar ratios of maleimide and 1-octene could be used, while in batch a 20-fold excess of 1-octene was employed. (Reprinted with permission from Ref. 49. Copyright © 2015 Elsevier.)

be exchanged by a maleimide after polymerization. Success of the polymerization as well as the maleimide introduction was proven via electrospray ionization mass spectrometry (ESI-MS).

The first step for a successful [2 + 2] cycloaddition is the excitation of the UV-absorbing ene to create the reactive intermediate which can react with an alkene. Efficiency in this step is dramatically increased by switching from conventional batch reactors to a photoflow system. With this change in reactors, the reaction time was reduced from days to minutes (see Figure 8.12 for a comparison of the small-molecule counterpart of the polymer modification). Additionally to the significant reduction in reaction time, also a reduction of the ene:enone ratio was achieved. In batch, a 20:1 ene:enone ratio is needed to reach quantitative conversion of the starting material. In flow it was reduced to equimolar ratio for reaction between two low molecular weight compounds. When switching to the polymer maleimide, addition of 10 mol% of a photosensitizer was required, which is, however, accompanied by a further rate increase (quantitative conversion within 1 min).

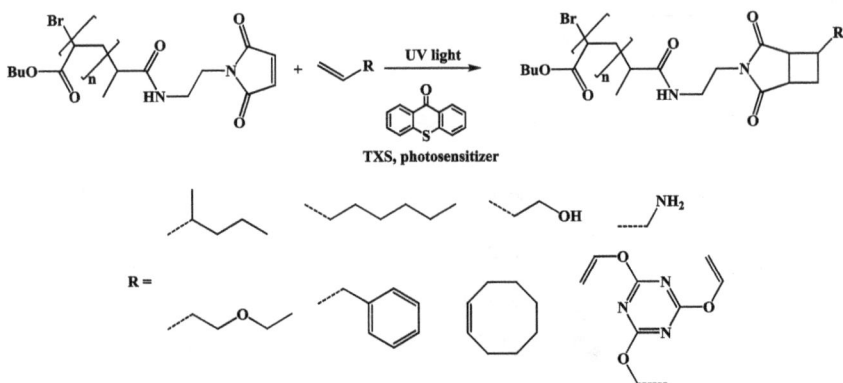

Scheme 8.3. Polymer endgroup modification in photoflow via [2 + 2] alkene–enone cycloaddition in the presence of the photosensitizer thioxanthone.

Via this [2 + 2] polymer modification, various functional alkenes were introduced into the polymer chain such as alcohols, ethers and multifunctional allyl components. Good yields were achieved in all cases (Scheme 8.3). In summary, the efficient and fast [2+2] cycloaddition can be a useful tool for polymer modification and introduction of functionalities when being performed in flow, which is remarkable keeping the low efficiency of the associated batch processes in mind.

8.4 Microparticles from UV-Photoflow Polymerization

As viscosity is a severe obstacle in solution polymerization, polymer synthesis is largely limited to low to intermediate molecular weights when carried out in homogenous solution. Only in very high dilution, high molecular weights are achievable in continuous flow. By changing the reaction mode to droplet-based or emulsified systems, high-molecular weight materials become available as viscosity is in such cases not directly correlated with molecular weight anymore. For such flow systems, also photoinduced polymerizations are of high interest, as they allow to separate the droplet formation and/or the nucleation process from the polymerization itself. In this section, the synthesis of polymer particles will be outlined, followed by approaches on inorganic nanoparticles, where the same advantage is utilized.

Different research groups have reported the synthesis of microparticles in microfluidic chip reactors via photoinitiated polymerization. Yeh and Lin described how to make very uniform microparticles via a water-in-oil emulsion flow technique.[50] With their developed microfluidic chip in combination with a 365-nm UV-light source they produced microparticles at low cost and high throughput. By varying the mineral oil flow rate, they are able to control the particle size between 75 and 300 μm diameter with narrow size distribution. Lee and coworkers present in their work a way to introduce a biocatalyst into polymeric microparticles. Therefore, they also use an emulsion microfluidic system in combination with a UV-light source.[51] Hwang *et al.* encapsulated magnetic nanoparticles in hydrogel microparticles.[52] To reach this aim, they also used a water-in-oil emulsion in a T-junction microfluidic device. By building microfluidic devices with different microchannels, the geometry of the particles can be influenced. After the monodisperse magnetic emulsion droplets were created in the T-junction, the morphology of the particles was locked-in via photoinitiated polymerization of the droplets. As geometries spheres, disks and plugs were achieved. The obtained microparticles showed superparamagnetic behavior. An earlier work of the same group described how the microparticle geometry can be influenced by the microchannel geometry at the light source.[53] They also used a similar setup to make multifunctional superparamagnetic Janus particles.[54] For this, they mix magnetic and non-magnetic prepolymer solution as water phase in mineral oil and polymerize in flow via photoinitiation. Under an external magnetic field, the resulting Janus spheres can self-assemble in stable chainlike microstructures. Other Janus particles via photoinitiated polymerization in emulsion-flow process were made by Chen *et al.*[55] and Nie *et al.*[56] Microfluidic emulsion technology was also used in combination with photoinitiated polymerization to create microcapsules.[57] To achieve this, a double-emulsion system was employed which additionally us allowed to influence the mechanical properties of the microcapsules. The thickness of the shell could be influenced by varying the flowrates of the inner and intermediate phases.

Du Prez and coworkers showed in their work the production of different polymer beads in continuous-photoflow reactors.[58-61] They generated droplets in a water-in-oil high internal phase emulsion. After the photopolymerization of the monomer droplets, particles with large macropores were obtained. Besides small spherical polymer beads of less than 400 μm diameter also rods.[58] and capsules.[60] could be made. In further reports they used thiol-ene and thiol-yne chemistry in similar reaction setups to produce monodisperse macroporous and noporous functional beads. After the fast radical mediated polymerization, no further functionalization steps were required to achieve the targeted functional polymer beads.[59]

8.5 Inorganic Materials from Flow Process

Not only polymer materials are available from photoflow reactions, but also the synthesis of inorganic nanoparticles is accessible in such reactors. The synthesis of silver, gold, platinum, palladium, rhodium and iridium nanoparticles has been reported as well as the synthesis of hybrid polyacrylamide/silver composite particles. The high interest in academic research and industry in noble metal nanoparticles is caused by their catalytic activities [62-64] and plasmonic properties.[65-67] These properties depend on the materials itself as well as on the size and shape of the nanoparticles. In the past, different pathways have been developed to synthesize small particles with high surface-to-volume ratios. Besides nanocrystals also non-spherical particles like nanorods, nanodisks and nanoprisms were made.[68] To create these specific particles, synthesis procedures with high reproducibility and yield quality such as narrow particle dispersities are required.[69] Batch processes often exhibit concentration and temperature gradients which lead to inhomogeneous reaction conditions and results in quite large dispersities.[68] Different research groups have proven that the employment of microflow approaches lead to a major improvement of noble metal nanoparticle particle homogeneity.[70-76] The short and good mixing conditions in microflow reactors allow for homogeneous initiation of nucleation and thus for narrow size distributions.

8.5.1 *Platinum group nanoparticles*

Hafermann and Köhler decouple in their work the synthesis of noble metal nanoparticles into four partial processes[68,69]:

(i) mixing of reactant solution
(ii) initiation of nucleation
(iii) termination of nucleation
(iv) completion of particle growth

The process separation in microflow synthesis allows to work under optimal reaction conditions for each step to obtain high-quality yields. In batch reactions all steps are performed in parallel as they are hard to separate. Process separation itself is achieved by employing a photochemical microcontinuous-flow process (Figure 8.13).

The reactor was build out of three individually controlled syringe pumps, poly-(tetrafluoroethylene) (PTFE) tubing (0.5 mm inner diameter), 4-port-manifold and FEP tubing for the photochemical initiation section. In the 4-port manifold, 1-mM aqueous noble salt solution, perfluoromethyldecalin and a 1:1 mixture of 3-mM 2-hydroxy-4'-(2-hydroxyethoxy)-2-methylpropiophenone (HMP) and 2% poly(vinylpyrrolidone) (PVP) are mixed. The addition of the water-immiscible perfluoromethyldacalin leads to the creation of a microsegmented flow. Through a higher flow rate of the carrier

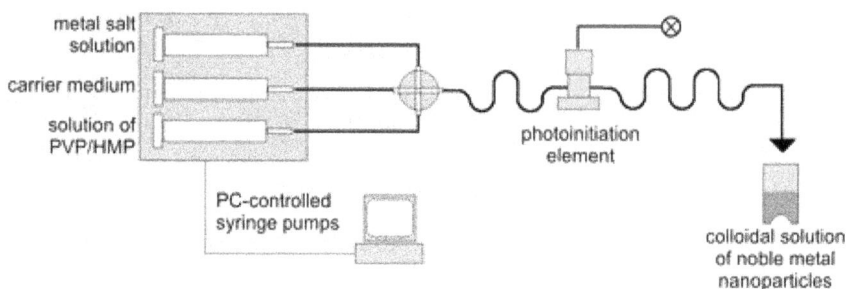

Fig. 8.13. Experimental setup for synthesis of Au, Pt, Pd, Ir and Rh nanoparticles in a three-step photochemical microcontinuous-flow process. (Reprinted with permission from Ref. 68. Copyright © 2015 John Wiley & Sons.)

medium compared to the aqueous phase, a regular formation of segments was achieved. For mixing, a 10-cm knotted PTFE tube was attached directly to the manifold. Followed by FEP tubing (5 cm) for the photoinitiation and a 50-cm knotted PTFE part as residence loop. Start and termination of the photoinitiation is defined by the length of the irradiation zone, here 0.2 cm.

Noble metal nanoparticles, in specific gold,[69] iridium,[68] palladium,[68] platinum[68] and rhodium[68] were synthesized at total flow rates between 43.75 and 700 μL min^{-1}. Through a change of the total flow rate, the nanoparticle size can be influenced. There is no general trend visible if slow or fast flow rates are needed to obtain the smallest nanoparticles. For example, to obtain gold nanoparticles of 2.5 nm in diameter, a total flow rate of 87.5 μL min^{-1} needs to be applied. However, to obtain platinum particles of the same size, a total flow rate of 700 μL min^{-1} gives the best yield with lowest dispersity.[68]

A comparison between batch and flow synthesis gold nanoparticles shows similar particle diameters but a significant lower size distribution. For the other used noble metals, a clear reduction in particle size in the flow process is observed (Table 8.3).

Table 8.3. Comparison of the size of noble metal nanoparticles synthesized via photochemical flow processes, thermal flow and conventional batch synthesis.

Metal	Precursor	Method/reagent	Particle sizes
Au	HAuCl$_4$	Batch	2.9 nm (Ref. 69)
		Photoflow/HMP	2–3 nm (Ref. 69)
Ir	H$_2$IrCl$_6$	Batch/Ethanol	5.8 nm (Ref. 77)
	IrCl$_3$	Photoflow/HMP	2.5 nm (Ref. 68)
Pd	Pd(acac)$_2$	Batch/Ligand exchange	8.7 nm (Ref. 78)
	Pd(NO$_3$)$_2$	Photoflow/HMP	2.5 nm (Ref. 68)
Pt	H$_2$PtCl$_6$	Batch/Ethanol	5.0 nm (Ref. 79)
		Flow	3.0 nm (Ref. 80)
		Photoflow/HMP	2.5 nm (Ref. 68)
Rh	RhCl$_3$	Batch/Ethylene glycol	3–7 nm (Ref. 78)
		Photoflow/HMP	2.5 nm (Ref. 68)

The obtained gold nanoparticles can be used as seed particles for the preparation of enlarged gold nanoparticles with defined sizes between 4 and 15-nm depending on the seed particle concentration. In contrast to other particle growing methods, no surfactants or shape-controlling additives are required by employing seeds from the photochemical micro-flow process.

8.5.2 *Silver nanoparticles*

Maggini and coworkers demonstrated how to grow silver nanoparticles (AgNP) under continuous-photoflow conditions.[81] In a first step, they generate via a photochemical process small (<10 nm) citrate-stabilized AgNP (seeds). The AgNP seeds were made in a flow reactor consisting out of a PTFE tubing coiled around an immersion-well reactor equipped with a high-pressure mercury lamp. A deoxygenated aqueous stock solution containing silver nitrate, sodium citrate and the photoinitiator Irgacure 2959 was pumped into the reactor via a syringe pump. The collected yellow samples were analyzed via UV–Vis spectroscopy. All samples of different flow rates show a surface plasmon band at 402 nm (Figure 8.14). With increasing irradiation time the lamp induces extensive aggregation of the nanoparticles which leads to the observed decrease of the absorption

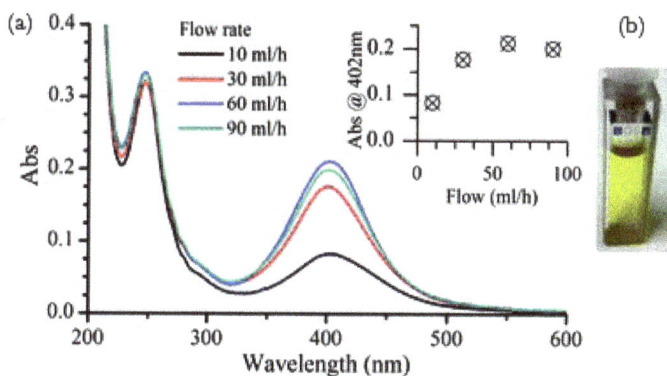

Fig. 8.14. UV–Vis absorption spectra (a) and color of the AgNP colloids (b). (Reprinted with permission from Ref. 81. Copyright © 2013 The Royal Society of Chemistry.)

at 402 nm. This change in particle size can also be observed with the eye by a color change of the collected samples from yellow to opalescent.

In a second step, growth of AgNP with different shapes is possible from these seeds.[82,83] through the so-called *photovoltage mechanism*.[84] This process requires the citrate which is adsorbed on the seed nanoparticles, oxygen and light. The adsorbed citrate is photooxidized, yielding acetone-1,3-dicarboxylate and CO_2. Electrons are injected into the particle with the decarboxylation. This induces the deposition and reduction of Ag^+ ions to silver.[81] The morphology of the nanoparticles is mainly influenced by the amounts of citrate and oxygen in the solution.[85,86] Till today this process is conventionally mainly performed in cuvettes with a limited volume. Due to this volume restrictions, the process is time consuming and lager amounts of AgNPs are difficult to access conventionally. Also reproducibility poses a problem in the batch-based production. Thus, also here continuous-flow reactors can be used to increase the productivity and the reproducibility of this process. To this end a microflow reactor with two inlets and a volume of 400 μL was made and equipped with an array of five LEDs (455, 505 and 627 nm). One inlet was used to load the reactor with an aqueous AgNP-seed solution which was oxygenated through the second inlet. The comparison of UV–Vis spectra of grown AgNP in flow and batch is indicating a better conversion of the starting seeds. This is notable especially since the irradiation time was only 20% of the exposure time in bulk. As side effect no larger, anisotropic particles as side products are formed.

8.5.3 *Polyacrylamide/silver composite particles*

Other possible applications of metal nanoparticles lay in polymer/metal composite materials. These materials feature special mechanical, electronic, optical, and chemical properties which have a growing interest for their applications in diverse fields.[87–91] Composite materials are mainly made by mixing of the components or through *in situ* formation by chemical synthesis.[92] One way to handle these composite materials is the synthesis of microparticles, which are synthesized by application of a droplet-based microfluidic technique with high homogeneity in size and chemical composition.[93,94] Therefore, preformed monomer droplets are polymerized via photoinitiated radical polymerization.[95] Polymer particles containing metal particles are of high interest due to the complementary nature of the chemical and physical properties of both materials.[96]

Such a process can be carried out in a two-step segmented flow process (Figure 8.15). In the beginning, the required AgNPs were

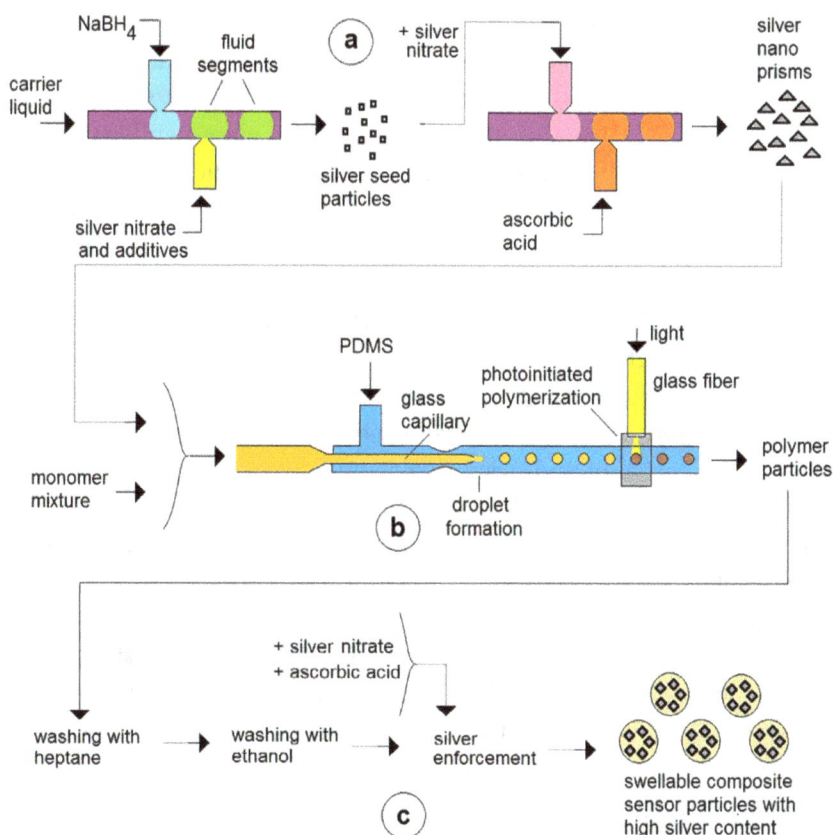

Fig. 8.15. Process strategy and principle of microfluidic arrangement for the synthesis of polyacrylamide/silver composite nanoparticles. (Reprinted with permission from Ref. 96. Copyright © 2013 American Chemical Society.)

synthesized as before with a segmented-flow technique. As a variation, the silver seed synthesis was carried out via a reduction of silver nitrate through sodium borohydride in aqueous solution within the presence of the sodium salt of polystyrenesulfonic acid. The segments were formed in a stream of perfluorinated alkane. The formation of silver nanoprisms was achieved by a silver-catalyzed silver deposition using ascorbic acid at room temperature. This step was also carried out in microsegmented flow and lead to a significant improvement in the homogeneity of nanoparticles.[97] For the composite particles, a premix of inorganic and organic components was made and followed by regular droplet formation in laminar flow in polydimethylsiloxane (PDMS). For stable flow conditions, with low pulsation, syringe pumps were employed. The monomer mixture droplets were formed regularly by pushing the mixture through a glass capillary (final opening 0.25 mm internal diameter) into the liquid PDMS jet. A droplet diameter of about 0.45 mm was reached with a monomer mixture flow rate of $20\,\mu L\ min^{-1}$ in $150\,\mu L\ min^{-1}$ PDMS. The droplet formation could be monitored with a microscopic video camera (Figure 8.16). Polymerization of the droplets was initiated by passing a UV light source with a residence of less than 1 s. After less than 10 s polymerization time, the solid particles could be collected. For purification, the particles (dispersed in PDMS) were washed with heptane and ethanol and dried in air. The droplet size could be controlled via the flow rates or the diameter of the glass capillary. With this setup, droplet sizes between 0.7 mm and $50\,\mu m$ were realized. The droplets with a diameter of 0.45 mm result in particles of around 0.3 mm after drying (Figure 8.17).

After the heptane and ethanol washing step, the particles were treated with a 0.1 M ascorbic acid and 10 mM silver nitrate solutions. This lead to a deposition of silver on the surface a few seconds after mixing. Similar reactions with nanoparticles without incorporated silver particles need an increase in temperature for fast silver deposition. The authors conclude that the particles surface catalyzes the silver deposition.

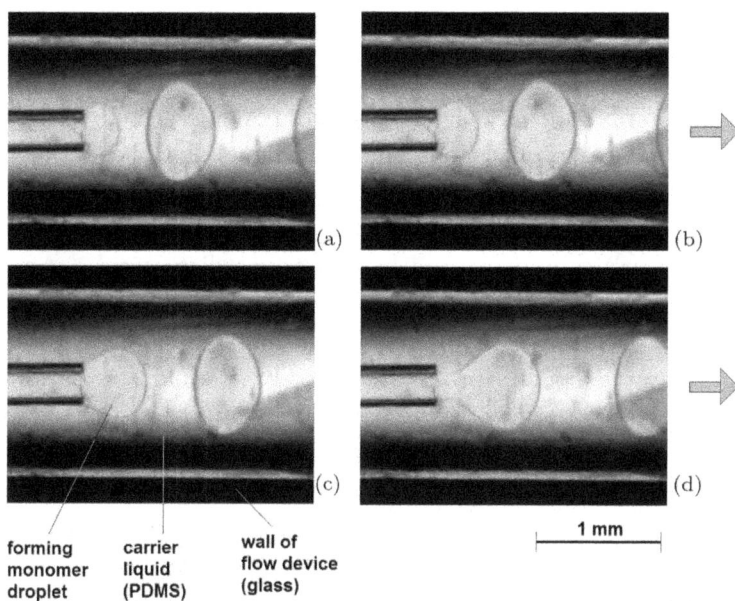

Fig. 8.16. Regular monomer mixture droplet formation in a stream of PDMS carrier flow. (Reprinted with permission from Ref. 96. Copyright © 2013 American Chemical Society.)

Fig. 8.17. Optical image of a group of polymer/silver composite particle (a) and SEM image of the particle surface (b). (Reprinted with permission from Ref. 96. Copyright © 2013 American Chemical Society.)

8.6 Conclusion

Photoflow synthesis opens the doors to large-scale reproducible production of a broad variety of materials. As described above, the combination of photoreactions and continuous-flow techniques have given access to very precise materials synthesis methodologies. Products that have been available only in small quantities — often containing impurities or featuring high product dispersities — can be easily upscaled when using photoflow. From complex multiblock copolymers over polymer microbeads to metal nanoparticles, a significant improvement in product quality, time-yield correlation and reproducibility is observed, whereby the advantage of the photoflow process stems in practically all cases from the high light efficiency reached in these reactors in combination with the high temporospatial control that can be used to trigger certain reactions only in distinct compartments of the reactor. While many of the presented procedures are yet just beyond a proof-of-concept stage, the vast potential is clearly visible and a steep increase in photoflow processes in materials design can be envisaged. Upscaling of photochemical material synthesis has to date been a severe obstacle, which can now be solved and hence give access to a multitude of applications in which these materials can in future be employed.

References

1. G. Gody, T. Maschmeyer, P. B. Zetterlund and S. Perrier, *Nat. Commun.* **4**, 2505 (2013).
2. J. Chiefari, Y. K. Chong, F. Ercole, J. Krstina, J. Jeffery, T. P. T. Le, R. T. A. Mayadunne, G. F. Meijs, C. L. Moad, G. Moad, E. Rizzardo and S. H. Thang, *Macromolecules* **31**(16), 5559 (1998).
3. K. Matyjaszewski and J. H. Xia, *Chem. Rev.* **101**(9), 2921 (2001).
4. K. Matyjaszewski, *Advances in Controlled/living Radical Polymerization* (American Chemical Society, 2003).
5. J. Nicolas, Y. Guillaneuf, C. Lefay, D. Bertin, D. Gigmes and B. Charleux, *Prog. Polym. Sci.* **38**(1), 63 (2013).
6. D. Benoit, V. Chaplinski, R. Braslau and C. J. Hawker, *J. Am. Chem. Soc.* **121**(16), 3904 (1999).
7. H. C. Kolb, M. G. Finn and K. B. Sharpless, *Angew. Chem. Int. Ed.* **40**(11), 2004 (2001).

8. J. A. Burns, C. Houben, A. Anastasaki, C. Waldron, A. A. Lapkin and D. M. Haddleton, *Polym. Chem.* **4**(17), 4809 (2013).
9. B. M. Rosen and V. Percec, *Chem. Rev.* **109**(11), 5069 (2009).
10. P. Derboven, P. H. Van Steen Berge, J. Vandenbergh, M.-F. Reyniers, T. Junkers, D. R. D'hoogei and G. B. Marin, *Macromol. Rapid Commun.* **36**, 2149 (2015).
11. T. Fukuyama, Y. Kajihara, I. Ryu and A. Studer, *Synthesis* **44**(16), 2555 (2012).
12. C. H. Hornung, C. Guerrero-Sanchez, M. Brasholz, S. Saubern, J. Chiefari, G. Moad, E. Rizzardo and S. H. Thang, *Org. Process Res. Dev.* **15**(3), 593 (2011).
13. N. Chan, M. F. Cunningham and R. A. Hutchinson, *J. Polym. Sci., A Polym. Chem.* **51**(15), 3081 (2013).
14. T. Noda, A. J. Grice, M. E. Levere and D. M. Haddleton, *Eur. Polym. J.* **43**(6), 2321 (2007).
15. B. Wenn, M. Conradi, A. D. Carreiras, D. M. Haddleton and T. Junkers, *Polym. Chem.* **5**(8), 3053 (2014).
16. Y. Shen and S. Zhu, *AlChE J.* **48**(11), 2609 (2002).
17. A. Melker, B. P. Fors, C. J. Hawker and J. E. Poelma, *J. Polym. Sci., A Polym. Chem.* **53**(23), 2693 (2015).
18. E. Baeten, B. Verbraeken, R. Hoogenboom and T. Junkers, *Chem. Commun.* **51**(58), 11701 (2015).
19. J. Haven, J. Vandenbergh, R. Kurita, J. Gruber and T. Junkers, *Polym. Chem.* **6**(31), 5752 (2015).
20. M. Conradi and T. Junkers, *J. Photochem. Photobiol. A* **259**, 41 (2013).
21. M. Conradi and T. Junkers, *Macromolecules* **47**(16), 5578 (2014).
22. J. P. Knowles, L. D. Elliott and K. I. Booker-Milburn, *Beilstein J. Org. Chem.* **8**, 2025 (2012).
23. A. Kermagoret, B. Wenn, A. Debuigne, C. Jerome, T. Junkers and C. Detrembleur, *Polym. Chem.* **6**(20), 3847 (2015).
24. M. Chen and J. A. Johnson, *Chem. Commun.* **51**(31), 6742 (2015).
25. Y.-M. Chuang, B. Wenn, S. Gielen, A. Ethirajan and T. Junkers, *Polym. Chem.* **6**(36), 6488 (2015).
26. C. Tonhauser, A. Natalello, H. Löwe and H. Frey, *Macromolecules* **45**(24), 9551 (2012).
27. D. Wilms, J. Klos and H. Frey, *Macromol. Chem. Phys.* **209**(4), 343 (2008).
28. J. Vandenbergh, T. de Moraes Ogawa and T. Junkers, *J. Polym. Sci., A Polym. Chem.* **51**(11), 2366 (2013).
29. J. Vandenbergh and T. Junkers, *Polym. Chem.* **3**(10), 2739 (2012).
30. N. Zaquen, E. Baeten, J. Vandenbergh, L. Lutsen, D. Vanderzande and T. Junkers, *Chem. Eng. Technol.* **38**(10), 1749 (2015).
31. M. Janata, L. Lochmann, P. Vlček, J. Dybal and A. H. E. Müller, *Die Makromolekulare Chemie* **193**(1), 101 (1992).
32. K. Iida, T. Q. Chastek, K. L. Beers, K. A. Cavicchi, J. Chun and M. J. Fasolka, *Lab Chip* **9**(2), 339 (2009).

33. J. J. Haven, J. Vandenbergh and T. Junkers, *Chem. Commun.* **51**(22), 4611 (2015).
34. J. Vandenbergh, T. Tura, E. Baeten and T. Junkers, *J. Polym. Sci. A Polym. Chem.* **52**(9), 1263 (2014).
35. A. Nagaki, Y. Takahashi, K. Akahori and J.-I. Yoshida, *Macromol. React. Eng.* **6**(11), 467 (2012).
36. T. Junkers and B. Wenn, *React. Chem. Eng.* **1**(1), 60 (2016).
37. D. Cambié, C. Bottecchia, N. J. W. Straathof, V. Hessel and T. Noël, *Chem. Rev.* **116**(17), 10276 (2016).
38. M. Oelgemöller and O. Shvydkiv, *Molecules* **16**(9), 7522 (2011).
39. B. D. A. Hook, W. Dohle, P. R. Hirst, M. Pickworth, M. B. Berry and K. I. Booker-Milburn, *J. Org. Chem.* **70** (19), 7558 (2005).
40. L. D. Elliott, J. P. Knowles, P. J. Koovits, K. G. Maskill, M. J. Ralph, G. Lejeune, L. J. Edwards, R. I. Robinson, I. R. Clemens, B. Cox, D. D. Pascoe, G. Koch, M. Eberle, M. B. Berry and K. I. Booker-Milburn, *Chem. Eur. J.* **20**(46), 15226 (2014).
41. E. E. Coyle and M. Oelgemoller, *Photochem. Photobiol. Sci.* **7**(11), 1313 (2008).
42. M. Oelgemoeller, *Chem. Eng. Technol.* **35**(7), 1144 (2012).
43. A. Anastasaki, V. Nikolaou, Q. Zhang, J. Burns, S. R. Samanta, C. Waldron, A. J. Haddleton, R. McHale, D. Fox, V. Percec, P. Wilson and D. M. Haddleton, *J. Am. Chem. Soc.* **136**(3), 1141 (2014).
44. M. Schappacher and A. Deffieux, *Macromolecules* **25**(25), 6744 (1992).
45. S. Jacob, I. Majoros and J. P. Kennedy, *Macromolecules* **29**(27), 8631 (1996).
46. B. Wenn, A. Martens, Y. Chuang, J. Gruber and T. Junkers, *Polym. Chem.* **7**, 2720 (2016).
47. C. Detrembleur, D.-L. Versace, Y. Piette, M. Hurtgen, C. Jerome, J. Lalevee and A. Debuigne, *Polym. Chem.* **3**(7), 1856 (2012).
48. D. Jérémy, K. Anthony, J. Christine, D. Christophe, D. Antoine, *Controlled Radical Polymerization: Materials*, Vol. 1188 (American Chemical Society, 2015), pp. 47–61.
49. T. Junkers, *Eur. Polym. J.* **62**, 273 (2015).
50. C.-H. Yeh and Y.-C. Lin, *Microfluid Nanofluid* **6**(2), 277 (2009).
51. W. J. Jeong, J. Y. Kim, J. Choo, E. K. Lee, C. S. Han, Beebe, D. J. G. H. Seong and S. H. Lee, *Langmuir* **21**(9), 3738 (2005).
52. D. K. Hwang, D. Dendukuri and P. S. Doyle, *Lab Chip* **8**(10), 1640 (2008).
53. D. Dendukuri, K. Tsoi, T. A. Hatton and P. S. Doyle, *Langmuir* **21**(6), 2113 (2005).
54. K. P. Yuet, D. K. Hwang, R. Haghgooie and P. S. Doyle, *Langmuir* **26**(6), 4281 (2010).
55. C.-H. Chen, R. K. Shah, A. R. Abate and D. A. Weitz, *Langmuir* **25**(8), 4320 (2009).
56. Z. Nie, W. Li, M. Seo, S. Xu and E. Kumacheva, *J. Am. Chem. Soc.* **128**(29), 9408 (2006).

57. Y. Hennequin, N. Pannacci, C. P. de Torres, G. Tetradis-Meris, S. Chapuliot, E. Bouchaud and P. Tabeling, *Langmuir* **25**(14), 7857 (2009).
58. M. T. Gokmen, W. Van Camp, P. J. Colver, S. A. F. Bon and F. E. Du Prez, *Macromolecules* **42**(23), 9289 (2009).
59. R. A. Prasath, M. T. Gokmen, P. Espeel and F. E. Du Prez, *Polym. Chem.* **1**(5), 685 (2010).
60. M. T. Gokmen, B. Dereli, B. G. De Geest and F. E. Du Prez, *Part. Part. Syst. Char.* **30**(5), 438 (2013).
61. M. T. Gokmen, B. G. De Geest, W. E. Hennink and F. E. Du Prez, *ACS Appl. Mater. Inter.* **1**(6), 1196 (2009).
62. M. Moreno-Mañas, R. Pleixats, *Acc. Chem. Res.* **36**(8), 638 (2003).
63. P. Mukherjee, C. R. Patra, A. Ghosh, R. Kumar and M. Sastry, *Chem. Mater.* **14**(4), 1678 (2002).
64. R. M. Crooks, M. Zhao, L. Sun, V. Chechik and L. K. Yeung, *Acc. Chem. Res.* **34**(3), 181 (2001).
65. A. Knauer, S. Schneider, F. Möller, A. Csáki, W, Fritzsche and J. M. Köhler, *Chem. Eng. J.* **227**, 80 (2013).
66. S. Link and M. A. El-Sayed, *J. Phys. Chem. B* **103**(21), 4212 (1999).
67. G. Schmid and L. F. Chi, *Adv. Mater.* **10**(7), 515 (1998).
68. L. Hafermann and J. M. Köhler, *Chem. Eng. Technol.* **38**(7), 1138 (2015).
69. L. Hafermann and J. M. Köhler, *J. Nanopart. Res.* **17**(2), 1 (2015).
70. A. Knauer and J. M. Koehler, *Nanotechnol. Rev.* **3**, 5 (2014).
71. M. Wörner, *Microfluid Nanofluid* **12**(6), 841 (2012).
72. A. Knauer, A. Csáki, F. Möller, C. Hühn, W. Fritzsche and J. M. Köhler, *J. Phys. Chem. C* **116**(16), 9251 (2012).
73. Z. Wan, W. Luan and S.-T. Tu, *J. Phys. Chem. C* **115**(5), 1569 (2011).
74. A. Knauer, A. Thete, S. Li, H. Romanus, A. Csáki, W. Fritzsche and J. M. Köhler, *Chem. Eng. J.* **166**(3), 1164 (2011).
75. A. B. Theberge, F. Courtois, Y. Schaerli, M. Fischlechner, C. Abell, F. Hollfelder and W. T. S. Huck, *Angew. Chem. Int. Ed.* **49**(34), 5846 (2010).
76. J. Wagner and J. M. Köhler, *Nano Lett.* **5**(4), 685 (2005).
77. Y. Tonbul, M. Zahmakiran and S. Özkar, *Appl. Catal. B Environ.* **148–149**, 466 (2014).
78. S. M. Humphrey, M. E. Grass, S. E. Habas, K. Niesz, G. A. Somorjai and T. D. Tilley, *Nano Lett.* **7**(3), 785 (2007).
79. M. Liu, J. Zhang, J. Liu and W. W. Yu, *J. Catal.* **278**(1), 1 (2011).
80. J. C. Scaiano, K. G. Stamplecoskie and G. L. Hallett-Tapley, *Chem. Commun.* **48**(40), 4798 (2012).
81. S. Silvestrini, T. Carofiglio and M. Maggini, *Chem. Commun.* **49**(1), 84 (2013).
82. S. Chen and D. L. Carroll, *Nano Lett.* **2**(9), 1003 (2002).
83. R. Jin, Y. Cao, C. A. Mirkin, K. L. Kelly, G. C. Schatz and J. G. Zheng, *Science* **294**(5548), 1901 (2001).
84. X. Wu, P. L. Redmond, H. Liu, Y. Chen, M. Steigerwald and L. Brus, *J. Am. Chem. Soc.* **130**(29), 9500 (2008).

85. C. Xue, G. S. Métraux, J. E. Millstone and C. A. Mirkin, *J. Am. Chem. Soc.* **130**(26), 8337 (2008).
86. M. Maillard, P. Huang and L. Brus, *Nano Lett.* **3**(11), 1611 (2003).
87. B. Adhikari and S. Majumdar, *Prog. Polym. Sci.* **29**, 699 (2004).
88. R. F. Gibson, *Composite Struct.* **92**, 2793 (2010).
89. D. W. Hatchett and M. Josowicz, *Chem. Rev.* **108**, 746 (2008).
90. Y. Ofir, B. Samanta and V. M. Rotello, *Chem. Soc. Rev.* **37**, 1814 (2008).
91. J. F. Tressler, S. Alkoy, A. Dogan and R. E. Newnham, *Composites A* **4**, 477 (1999).
92. K. S. Iyer, B. Zdyrko, S. Malynych, G. Chumanov and I. Luzinov, *Soft Matter* **7**, 2538 (2011).
93. J. I. Park, A. Saffari, S. Kumar, A. Guenther and E. Kumacheva, *Ann. Rev. Mater. Res.* **40**, 415 (2010).
94. S. Q. Xu, Z. H. Nie, M. Seo, P. Lewis, E. Kumacheva, H. A. Stone, P. Garstecki, D. B. Weibel, I. Gitlin and G. M. Whitesides, *Angew. Chem. Int. Ed.* **44**, 724 (2005).
95. C. A. Serra and Z. Chang, *Chem. Eng. Technol.* **31**, 1099 (2008).
96. J. M. Köhler, A. März, J. Popp, A. Knauer, I. Kraus, J. Faerber and C. Serra, *Anal. Chem.* **85**(1), 313 (2013).
97. A. Knauer, A. Csaki, F. Möller, C. Hühn, W. Fritzsche and J. M. Köhler, *J. Phys. Chem. C* **116**, 9251 (2012).

Chapter 9

Industrial Photochemistry: From Laboratory Scale to Industrial Scale

Timothy Noël,[*,†] Marc Escriba Gelonch[*] and Kevin Huvaere[†]

Department of Chemical Engineering & Chemistry
Laboratory for Micro Flow Chemistry & Process Technology
Eindhoven University of Technology, De Rondom 70
5612AP Eindhoven, The Netherlands

†EcoSynth, Plassendale Chemistry Site, Stationsstraat 123
8400 Ostend, Belgium
‡t.noel@tue.nl

9.1 Introduction

Photochemistry is probably one of the oldest activation modes to enable organic synthetic reactions.[1] Despite this fact, the use of photochemistry has been neglected for many decades. This can be mainly attributed to the bad reputation of photochemistry which arises from its limited scalability and low selectivity of the reactions.[2,3] Arguably, many chemists have claimed that there is no future for photochemistry in both academia and industry.

However, photochemistry has witnessed a tremendous revival in the last decade. In our opinion, this renewed interest can be attributed to two main aspects. First, the emergence of visible-light photoredox catalysis as a new activation mode allowed to facilitate reaction pathways which were previously elusive. Although this field exists already from the early 1980s,[4] it was the group of MacMillan who showed in 2008 for the first time the tremendous power of this catalytic mode.[5] Their landmark paper proved to be the required spark that ignited the entire synthetic organic chemistry field.[6–10] Rapidly, a plethora of new applications were published by the community. Photoredox catalysis was particularly attractive for synthetic chemists as it provided very mild and simple reaction conditions (room temperature, visible light, avoidance of hazardous reagents), high selectivities and new reactivities. The use of visible-light LEDs allowed to dramatically reduce the energy costs, as UV photochemistry is inherently expensive due to the high light source cost, their limited lifetime and high energy consumption. A second reason for the revival of photochemistry is the development and the gaining popularity of microreactor technology. Photochemistry has always suffered from limited scalability due to the Bouguer–Lambert–Beer limitation. Microreactors allow to scale photochemistry from small scale to large quantities.[11–14] Especially on a laboratory scale, continuous-flow microreactor technology offered the potential to prepare gram to kilogram quantities, making photochemistry an alternative protocol for sometimes lengthy reaction sequences.

In this chapter, we provide an overview of the different scaling strategies for photochemistry. We also give an overview of some important industrial processes in which photochemistry provides a key advantage compared to other synthetic schemes.

9.2 Scale-Up Strategies

One of the critical aspects of scaling photochemical reactions is the occurrence of photon transport limitations.[15] Consequently, a dimension enlarging strategy, typically used for scale-up of chemical reactions, cannot be employed. In such large diameter reactors, the irradiation does not penetrate the liquid efficiently and thus longer irradiation times need to be used, often leading to the formation of by-products.

With microreactors, essentially two strategies can be distinguished to scale photochemical reactions: (i) using a single microreactor device and operating it at higher flow rates and/or longer operation times, and (ii) numbering-up.[13]

The first strategy is very convenient as it allows to optimize the reaction conditions on a small scale and subsequently to use the same device to prepare larger quantities by continuous introduction of starting materials. The main advantage is that there is no time-consuming re-optimization of the reaction conditions necessary. Increasing the throughput is possible by increasing the flow rates while keeping the total residence time the same, i.e. by increasing the total volume of the reactor. The use of high flow rates has a pronounced influence on the hydrodynamics and the heat transfer, as both are increased substantially. This can lead to a further reduction of the reaction time. However, the higher the flow rate, the higher the pressure drop over the reactor becomes. As many photochemical reactions are performed in a photon-limited regime, it should be noted that such reactions are often substantially accelerated in microreactors due to the optimal irradiation of the reaction mixture. This process intensification effect leads to a boost of the reactor throughput. With a single device strategy, typically several milligrams to hundreds of grams can be prepared. This is mostly sufficient in academic laboratories or even in medicinal chemistry programs to prepare enough material to screen the properties of the new drug or material.

If more material is required, several reactors can be placed in parallel (numbering-up) (Figure 9.1). Both internal and external numbering-up strategies can be used. External numbering-up is technologically the easiest as it is essentially achieved by placing perfect copies of a single device in parallel. This means that reactor, pumps and process control units are multiplied until the desired amount of material can be produced. The main advantage of this strategy is the stand-alone operability of each unique device, which makes that reactor failure does not influence the other reactors. However, as the main cost of any photomicroreactor assembly are the pumps and the control units, this strategy is not always economically feasible. In contrast, internal numbering-up will exclusively multiply

(a)

(b)

(c)

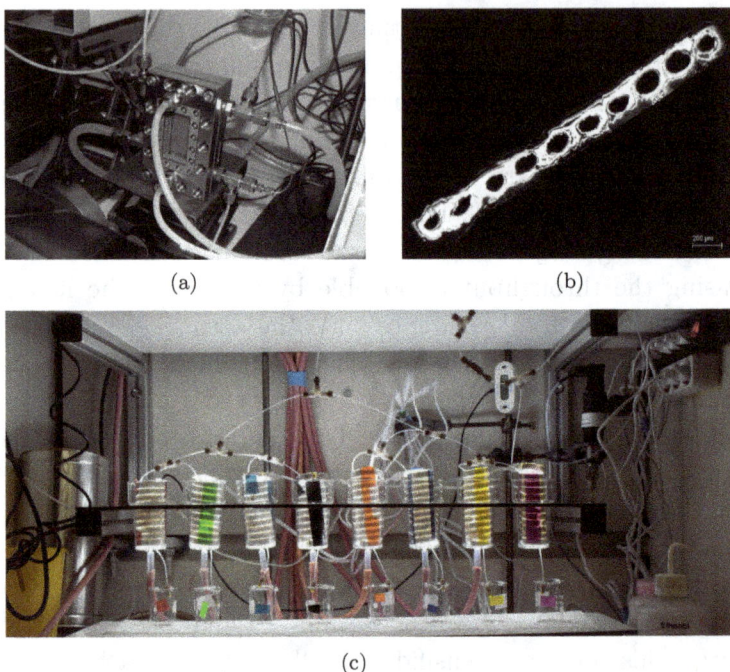

Fig. 9.1. Internal numbering-up strategies for photomicroreactors: (a) Falling-film microreactors. (Reprinted with permission from Ref. 9. Copyright © 2005 WILEY-VCH, Weinheim.) (b) Microcapillary films. (Reprinted with permission from Ref. 9. Copyright © 2013 American Chemical Society.) (c) Internal numbering-up of photomicroreactors for gas-liquid visible-light photoredox catalysis.

the reactor while the pumping system and process control units are shared. While this strategy is economical, it is challenging from an engineering standpoint the reagent streams have to be directed and equalized over the different photomicroreactors. Small differences in pressure drop will lead to a maldistribution resulting in non-optimal reaction conditions in the different reactors. Consequently, one of main parts in an internal numbering-up reactor assembly is the flow distributor section which has to ensure an equal mass flow throughput over the different reactors. Noël *et al.* have developed a modular internal numbering-up strategy to scale photocatalytic reactions with 2^n photomicroreactors ($n = 0, 1, 2, 3$) (Figure 9.1(c)).[16]

Commercially available T-mixers have been used as splitting devices. This branching tree structure splitting unit allowed to efficiently scale the photocatalytic aerobic oxidation of thiols to disulfides. Standard deviations <10% were observed indicating an excellent flow distribution potential. Other strategies for internal numbering up for photochemical reactions are the use of falling film microreactors (Figure 9.1(a))[17] and microcapillary films (Figure 9.1(b)).[18,19]

9.3 Solar Photochemical Reactors

The use of the sun as an eternal source of energy would represent a breakthrough for the chemical industry in terms of energy efficiency and sustainability. Since sunlight is often not intense enough, specialized equipment has been developed to concentrate the solar irradiation toward the reactor, e.g. sun tracking systems and refractors.[20] An example of an industrial-scale concentrating reactor is the PROPHIS facility close to Cologne in Germany (see Figure 9.2).[21] Glass reaction tubes (60 mm ID, 4.5 m length) are placed in the focal

Fig. 9.2. PROPHIS loop facility with parabolic mirrors allows to increase the intensity of the sun 30–32 times. (Reprinted with permission from Ref. 21. Copyright © 2005 Royal Society of Chemistry.)

Toray Caprolactam process

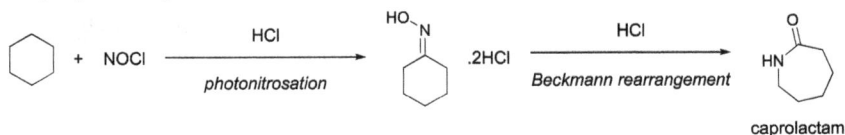

Scheme 9.1. Toray caprolactam process.

points of parabolic mirrors which allow to increase the photon flux up to 30–32 suns. The reaction volume that can be processed is 35–120 L providing a production of about 1 ton per year. Several reactions have been carried out in this device, including photooxidation reactions, photocyclizations and photo-Friedel–Crafts acylations.

Higher solar concentrations can be obtained in so-called solar furnaces (5000–20000 suns).[22] However, the high cost of such devices and the limited production scale reduces the application potential of solar devices for industrial applications.

On an industrial scale, solar energy is mostly used for water treatment applications.[23] Purification methods mostly rely on advanced oxidation processes where highly reactive oxygen species are used to degrade non-biodegradable organic compounds.[24]

9.4 Applications in Industrial Settings

9.4.1 *Synthesis of caprolactam: Toray process*

The Toray process is an industrial process to prepare caprolactam from cyclohexane (Scheme 9.1). Key in this process is a photonitrozation step which transforms cyclohexane into cyclohexanone oxime by reaction with nitrosyl chloride under light irradiation.[25] A subsequent Beckmann rearrangement results in the formation of caprolactam, which is used to manufacture nylon-6.

The photoreactor consists of a continuous reactor in which several mercury immersion lamps are positioned as shown in Figure 9.3.[17] The reaction mixture is kept below 20°C by a coolant, which is running through opaque channels, to minimize by-product formation. Cyclohexane is fed to the reactor and brought into contact with

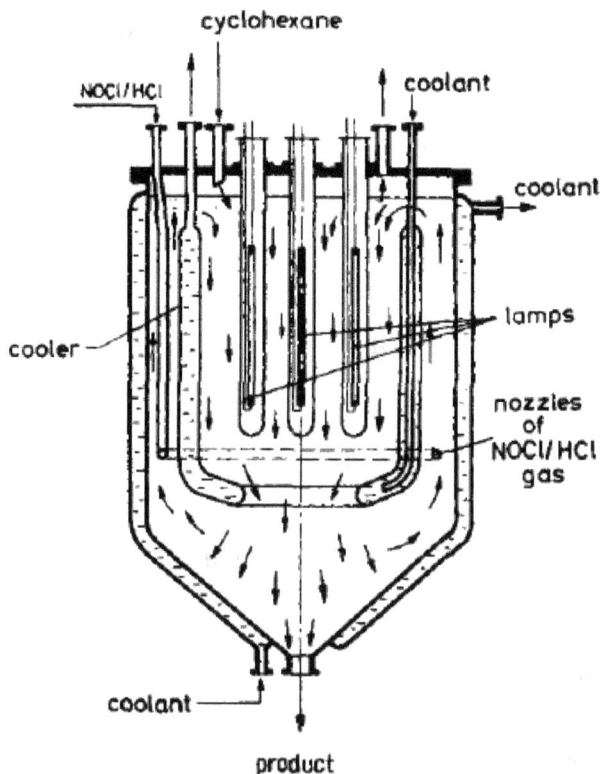

Fig. 9.3. Continuous photoreactor for the photonitrozation step in the Toray process. (Reprinted with permission from Ref. 26. Copyright © 2014 WILEY-VCH, Weinheim.)

gaseous NOCl and HCl. The cyclohexanone oxime is higher in density and forms fine droplets which coalesce and are removed from the bottom of the reactor.

One of the critical economic factors within any photochemical process is the energy cost. The Toray process requires large amounts of electricity and can only be economically viable in the presence of a cheap electricity source (e.g. nuclear power station or in modern times this will require cheap wind or solar energy). Furthermore, the photonitrozation is a radical chain process which makes that one photon gives rise to more equivalents of product (quantum yield >1). This provides a high efficiency of the light source.

The overall yield of the Toray caprolactam synthesis is 81% and is carried out on a 160,000-ton/year scale. The process uses cheaper starting materials than alternative caprolactam processes, however, the high overall energy costs makes that there exists only one photochemical caprolactam plant.

9.4.2 *Synthesis of vitamin D*

Vitamin D is a group of fat-soluble secosteroids responsible for the homeostasis of calcium and phosphorus, for cell proliferation, differentiation and apoptosis and for immune and hormonal regulation.[27] The current production of the "sunshine vitamin" mimics its synthesis *in vivo*. Using cholesterol obtained from skin of animals as a starting material, 7-dehydrocholesterol (7DHC) (provitamin D) is usually obtained through a four-step bromination–dehydrobromination process (50% yield). 7DHC is subsequently subjected to a photochemical process using UV-B irradiation (275–300 nm) (Scheme 9.2). Under these reaction conditions, a bond rearrangement takes place resulting in an electrocyclic ring opening of 7DHC to previtamin D. Also

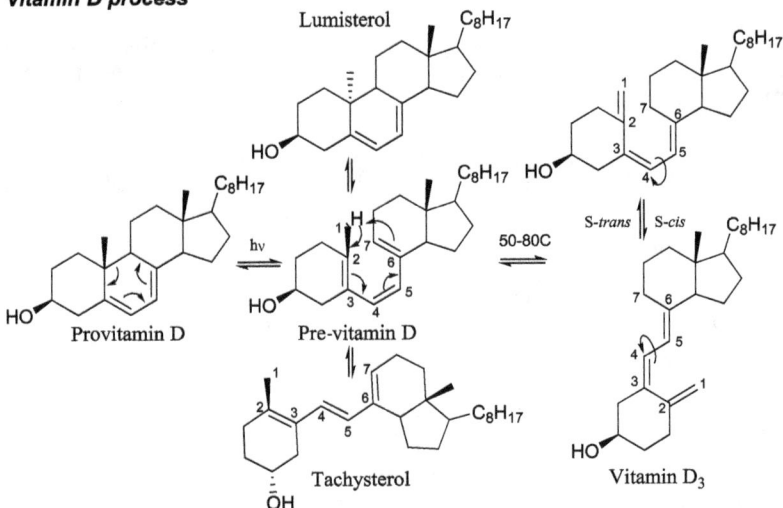

Scheme 9.2. Vitamin D photochemical process.

by-products are typically generated, such as lumisterol and tachysterol. The photochemical electrocyclic provitamin ring opening is followed by an antarafacial [1, 7] sigmatropic hydrogen rearrangement. Parameters like irradiation time, temperature and frequency of light play a relevant role in the final ratio of all isomers.

This complex photoreaction is carried out on an industrial scale in a quartz, temperature controlled and UV transparent reactor by using usually UV mercury medium-pressure lamps. The unspecific polychromatic radiation favors the formation of lumisterol and tachysterol as well, which are inactive and in some cases toxic. In order to minimize by-product formation, the process is stopped after low conversion of 7DHC. The unconverted 7DHC is recovered and send back to the photochemical reactor. A mixture of 7DHC dissolved in diethyl ether is commonly recirculated until the desired conversion is reached. Next, a solvent switch to methanol results into the precipitation of 7DHC which is removed by filtration. The liquid phase is subsequently concentrated *in vacuo* up to a pre-vitamin yellowish resin, which is finally heated to isomerize the pre-vitamin D into the cis-vitamin D3.

Vitamin D is a remarkable market at the moment. The global demand is divided in three between Europe, USA and the rest of the world. More than 80% of the production of vitamin D is destined for animal-feed purposes. Currently, the most important vitamin D producers are USA, Western Europe, India, China and Japan. The market is projected to reach about $2.5 billion by 2020 at a compound annual growth rate (CAGR) of 11%.[28]

9.4.3 *Synthesis of artemisinin: Sanofi process*

Artemisinin is one of most important antimalarials currently on the market.[29,30] Historically, artemisinin was isolated from the plant *Artemisia annua*. As a reaction on the unreliable agricultural supply, a mixed synthetic-biology route was developed with funding from the Bill and Melinda Gates Foundation (Scheme 9.3). In 2013, Sanofi launched a large-scale production facility to commercialize this process in Galessio (Italy).[31] This semi-synthetic process starts

Sanofi artemisinin process

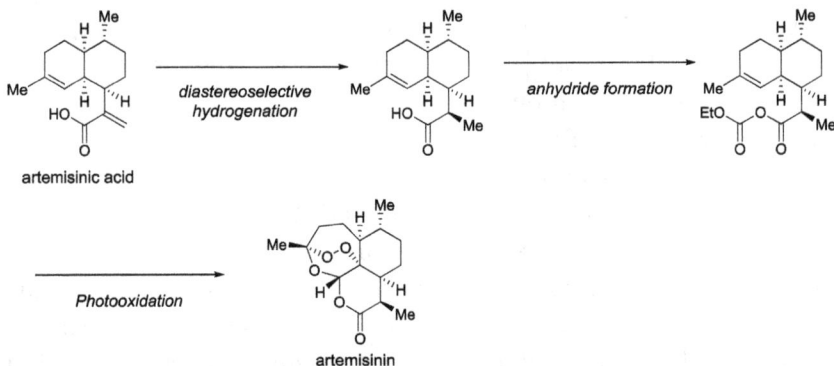

Scheme 9.3. Sanofi's artemisinin process with a photochemical singlet oxygen ene reaction as a key transformation.

from microbially derived artemisinic acid, which is first hydrogenated using a Ru-based enantioselective catalyst. Dihydroartemisinic acid is subsequently converted to a mixed anhydride, which is fed to a photochemical reactor. Herein, a singlet oxygen ene is carried out with tetraphenylporphyrin as a photocatalyst. A subsequent Hock cleavage, oxygenation and cyclization results in the formation of artemisinin in 55% overall yield (Scheme 9.4).

The photochemical reactor consists of a semi-batch reactor with a recirculation loop. A medium-pressure mercury/gallium lamp was used as photon source. The first pilot scale plant was launched in 2011 in Neuville and provided 50 kg batches (see Figure 9.4). Next, a large scale plant was built in Galessio which allowed to produce up to 60 ton of artemisinin in 2014 (in batches of 370 kg isolated artemisinin). This allows to cover one-third of the global need. Recent work from Sanofi demonstrated that a more energy-efficient reactor assembly can be obtained by using LEDs.[32] Despite the high costs of the LEDs, the total lifetime, maintenance and replacement costs made the use of LED irradiation economically feasible.

However, due to increased demand of artemisinin, more farmers cultivated the plant making the naturally derived artemisinin much cheaper ($250/kg compared to $350–400/kg for the synthetic

Fig. 9.4. Pilot-scale photochemical reactor for the synthesis of artemisinin. Note that the picture was taken when the reactor was under construction and the light losses were solved in the final assembly. (Courtesy of Sanofi.)

variant).[33] Recently, Sanofi has decided to sell its artemisinin plant in Garessio to Huvepharma. Huvepharma was already the producer of artemisinic acid and has now the entire production process under its control. Consequently, the company hopes to make the overall production process for semi-synthetic artemisinin cheaper and thus competitive with the naturally derived chemical.

9.4.4 *Synthesis of rose oxide: Dragoco process*

The industrial synthesis of rose oxide, with annual production of approximately 60–100 ton, is a benchmark example of a successfully-scaled photochemical reaction. As a mixture of four stereoisomers, synthetic rose oxide is a cheaper alternative to the natural fragrant oil extracted from Bulgarian rose for which 3000 kg of blossoms are needed to obtain 1 kg oil.[34]

The synthetic production, as developed by the German Dragoco company (currently Symrise), starts from β-citronellol in methanol

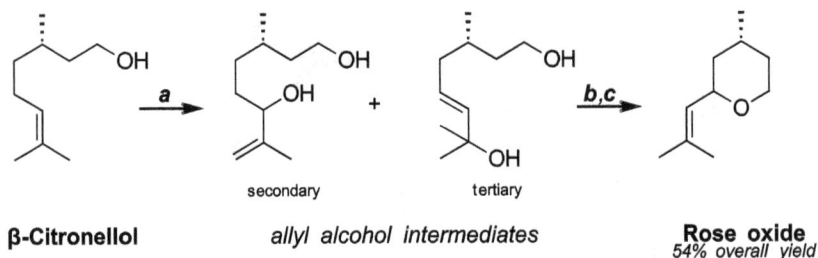

β-Citronellol   *allyl alcohol intermediates*   **Rose oxide**
                                                                   54% overall yield

Scheme 9.4. Rose Bengal-sensitized photooxidation to the intermediate allyl alcohols is key in the industrial synthesis of rose oxide from β-citronellol.

and involves a rose Bengal-sensitized, Schenck-type photooxygenation with formation of two hydroperoxide intermediates.[35] Irradiation using 5-kW mercury lamps is carried out in 3-m tall immersion-well reactors which, after reduction by sodium sulfite, gives a near quantitative conversion to the corresponding allyl alcohols in 3:2 ratio (tertiary to secondary).[36] Acid-catalyzed ring closure to the final pyran followed by steam distillation gives the desired rose oxide in 54% overall yield (Scheme 9.4).

Although the technology was developed over 50 years ago,[37,38] the process is assessed as particularly sustainable. According to life cycle assessment of different protocols, the Dragoco process outperforms alternative, more recent synthetic routes involving dark oxygenations.[39] Electrical power consumption was assigned as the main contributor to the environmental impact, a significant amount of which is consumed by the irradiation source. Moreover, it is worth noting that the latter suffers from relatively short lifetime and requires yearly replacement. In this respect, substituting sunlight for the artificial mercury arc as in a solar photoreactor is considered an attractive, sustainable option both in reducing power consumption and in eliminating the use of mercury-based irradiation sources.[40] According to calculations specific for rose oxide synthesis, avoiding these costs by using solar irradiation eventually lowers production cost by ~2.7% with further optimization feasible for solar production units situated in sunny regions.[41]

Next to sunlight, fast-paced development in LED technology may as well offer a substitute for mercury lamps. LEDs are low powered, have long lifetimes (∼10,000 h claimed) and may emit in a narrow bandwidth (hence optimized irradiation efficiency). Their compact design and modular organization in arrays make them especially useful for hybridization with continuous-flow microreactors. The small actual reactor volumes reduce hazards of working with oxygen-rich solvents as is the case in the rose oxide synthesis, while narrowbore, transparent tubes guarantee homogeneous irradiation conditions. Still, despite these benefits, rose oxide manufacturing in such process scores dramatically poor in the life cycle assessment in comparison to the Dragoco protocol.[39] Main culprit is the long irradiation time by a large series of lamps, thus emphasizing the need for careful investigation of energetic requirements prior to considering scale-up of photochemical syntheses.

9.4.5 *Photohalogenation*

Photohalogenation to introduce halogen atoms on a variety of substrates has developed into one of the few photoreactions that were scaled for the bulk chemical industry. Initiated by a light-induced homolytic cleavage of halogens, resulting free radicals trigger chain-carrying processes that outperform their thermally driven equivalent (Figure 9.5). Although efficiency of radical formation increases from fluorine to iodine, as a consequence of higher transition probability, basically only chlorine and bromine are useful for photohalogenation (see Scheme 9.5).[42]

   Photochlorination is most relevant as chlorine is readily fed in a gas–liquid reactor with the turbulence accounting for sufficient mass transfer to the liquid phase. Light sources typically include mercury lamps (few kW) positioned near a transparent tube reactor which gives converging irradiation inside the reaction mixture.[26] The tube diameter is a critical parameter when considering that a 10 cm path length through a 0.1% chlorine solution practically absorbs 90% of the light in the active range (300–500 nm). Alternatively, lamps are submerged in the reaction mixture when immersion-well setups are applied. Low-powered sources are preferred as high photon flux at the lamp surface may lead to excessive radical generation with

Initiation	$X_2 \xrightarrow{h\nu} 2X^\bullet$
Propagation	$RH + X^\bullet \longrightarrow R^\bullet + HX$
	$R^\bullet + X_2 \longrightarrow RX + X^\bullet$
Termination	$R^\bullet + X^\bullet \longrightarrow RX$
	$R^\bullet + R^\bullet \longrightarrow R_2$
	$X^\bullet + X^\bullet \longrightarrow X_2$

X	λ_{max} (nm)	ε (L mol^{-1}cm^{-1})	ΔH_D (X-X) (kJ/mol)	ΔH_F (H–X) (kJ/mol)
Cl	330	651	243	432
Br	420	16511	192	366

(a)   (b)

Fig. 9.5. (a) Initiation, propagation, and termination reactions involved in photohalogenation. (b) Photochemical properties of Cl$_2$ and Br$_2$, including wavelength of maximal absorption (λ_{max}) and the corresponding molar absorptivity (ε), and thermodynamical properties including heats of homolytic dissociation of the halogen reagent and hydrogen abstraction by the resulting radicals (ΔH_D and ΔH_F, respectively).

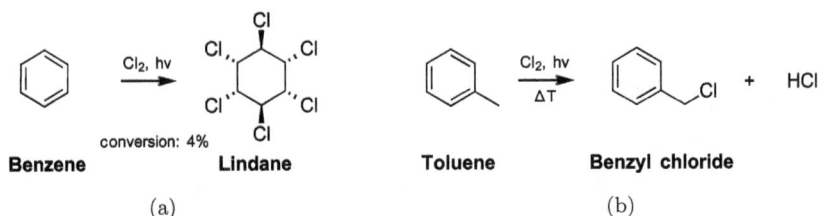

conversion: 4%

Benzene   **Lindane**   **Toluene**   **Benzyl chloride**

(a)   (b)

Scheme 9.5. (a) Photochlorination of benzene was used in the production of the γ-isomer of hexachlorocyclohexane, also known as lindane. Conversion of benzene was deliberately kept at or below 4% as to prevent clogging of the tube reactor by crystalline product. (b) Photochlorination of toluene in the production of benzyl chloride is carried out at increased temperature (80–110°C) to avoid aromatic chlorination.

secondary by-product formation. Still, both reactor types account for high quantum yields, typically well above 1000, which however inhibited efforts to optimize reactor systems. Fine tuning of reactions is typically obtained by controlling degree of conversion (e.g. via temperature control or limiting chlorine feed) rather than hardware modifications.

Long before concerns about toxicity and environmental persistence of chlorinated chemicals arose, light-induced addition of chlorine to benzene was an example process of industrial photochlorination.[43] A set of hexachlorinated cyclohexane stereoisomers was obtained, one of which being a potent pesticide known under the name lindane. Although the substance is now banned, hydrocarbons are still preferred substrates for photochlorination with exothermic hydrogen abstraction by chlorine radicals as the key propagating step. With an annual production of over 100,000 metric tons by, among others, Ineos and Lanxess, the synthesis of benzyl chloride from toluene is a very relevant case.[44] The reaction is preferably carried out near the boiling point of toluene, the conditions under which chlorination of the methyl side group (rather than the aromatic nucleus) are favored. Paraffins are generally chlorinated at lower temperatures (e.g. 30°C) with slight preference for substitution of tertiary hydrogens, although control is difficult particularly if larger hydrocarbons are reacted.[42]

Photobromination is generally more selective since bromine radicals are less reactive resulting in endothermic abstraction of (activated) hydrogens. In this respect, promising developments in continuous-flow microreactor technology have demonstrated that brominations of hydrocarbons are selectively and efficiently carried out using low-powered fluorescent or LED light sources.[45] Chlorination reactions in such systems require gas–liquid interfaces (e.g. in a falling film setup) and are carried out using other than chlorine reagents,[46,47] but further elaboration is pursued to assess their potential in industrial manufacturing. It is worth noting that even direct fluorination reactions, using specific reagents, could thus be considered.[48]

9.4.6 *Photoredox catalysis at Merck*

Researchers from Merck have developed in collaboration with the Knowles group a photocatalytic indoline dehydrogenation en route to elbasvir.[49] The photocatalytic method used an iridium-based photosensitizer and *tert*-butylperbenzoate as an environmentally benign oxidant (Scheme 9.6). No epimerization of the stereogenic hemiaminal center was observed under these reaction conditions. The reaction could be scaled to 100 g in a continuous-flow reactor constructed of

Indoline dehydrogenation

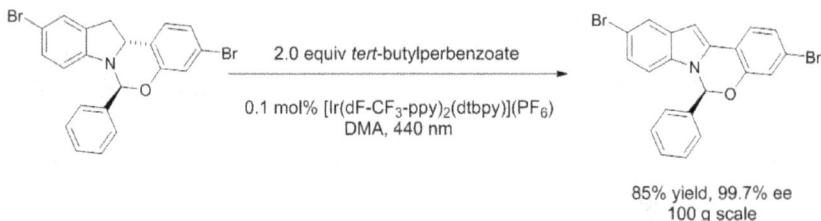

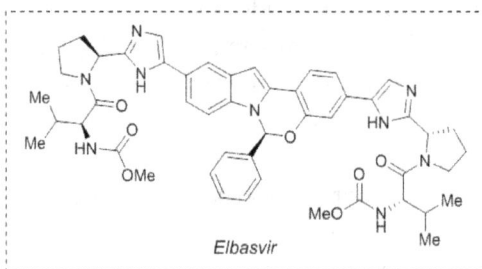

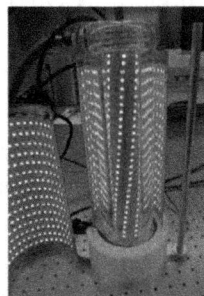

Scheme 9.6. Merck's continuous-flow photocatalytic indoline dehydrogenation for the synthesis of elbasvir, with a picture of the flow reactor.

PFA capillary tubing (3.2 mm ID, 15.24 m length, 150 mL volume) and blue LED irradiation (440 nm).

In collaboration with Britton *et al.*, researchers from Merck developed a direct photocatalytic C–H fluorination of (2*S*)-leucine (Scheme 9.7).[50] Decatungstate was used as the photocatalyst and *N*-fluorobenzenesulfonimide (NFSI) as the fluorination agent. The corresponding γ-fluoroleucine could be obtained on a larger scale by translating the protocol to continuous flow (PFA, 1.6 mm ID, 30.5 m length, 60 mL volume).

In order to increase the scale of their photocatalytic chemistry, Merck researchers developed a manufacturing-scale photoreactor, see Figure 9.6. The reactor consists of 440 m of capillary PFA tubing (3.2 mm ID, 3.5 L volume) which allowed to produce up to 2.7 kg/h of product. The irradiation efficiency was 24% corresponding with a 2.2-kW energy use.

Synthesis of γ-fluoroleucine

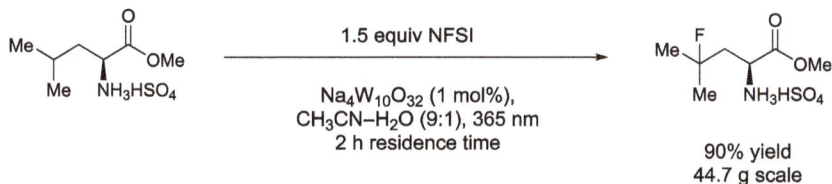

Scheme 9.7. Merck's photocatalytic C–H fluorination for the preparation of γ-fluoroleucine in continuous flow.

Fig. 9.6. Merck's manufacturing-scale continuous-flow photoreactor. (Courtesy of Merck.)

9.4.7 *Synthesis of 10-hydroxycamptothecin by Heraeus Noblelight*

An external numbering-up strategy was used by Heraeus Noblelight to scale-up the photochemical step in the synthesis of 10-hydroxycamptothecin and an ethylated derivative, which are precursors of the anticancer drugs irinotecan and topotecan (Scheme 9.8).[51]

Synthesis of 10-hydroxycamptothecin:

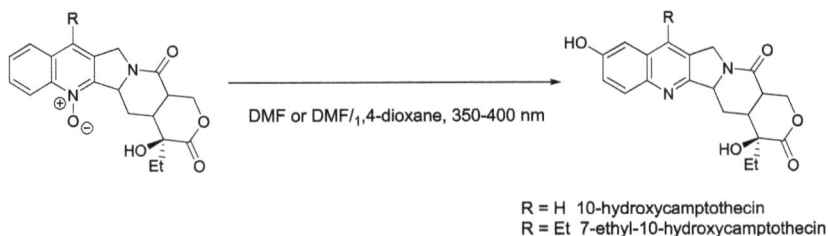

DMF or DMF/1,4-dioxane, 350–400 nm

R = H 10-hydroxycamptothecin
R = Et 7-ethyl-10-hydroxycamptothecin

Scheme 9.8. Photochemical synthesis of 10-hydroxycamptothecin and an ethylated derivative.

The system consists of 12 photomicroreactors in parallel and are individually irradiated with 350–400 nm light (from UV LEDs or high-pressure Hg lamps with appropriate spectral filters) (Figure 9.7). The reactor consists of two parallel quartz plates in which a thin film of the reaction solution (40–100 μm) is flowing. This photochemical production plant is able to produce about 2 kg/day of 10-hydroxycamptothecin (95% conversion, 90% yield). When compared to the corresponding batch reactor, a six-time higher dilution is required in batch and the conversion and yield are respectively 85% and 50%.

9.4.8 *Other examples from the pharmaceutical industry*

The pharmaceutical industry has been attracted by continuous-flow photochemistry as it enables scalability of photochemical reactions from laboratory scale to pilot/production scale. Some of the key examples of photochemistry in drug synthesis have been discussed before in greater detail, e.g. vitamin D_3 or artemisinin. Despite that many more pharmaceutical companies are interested in photochemistry, little to no details are typically divulged about their research activities. The information gathered here is based on what was disseminated in literature or on personal accounts; the examples are often results obtained on a laboratory scale.

Abbott Laboratories (Current AbbVie) have developed a lab-scale photomicroreactor (LOPHTOR) and evaluated its efficiency

Fig. 9.7. External numbering-up of a photochemical production unit by Heraeus Noblelight GmbH. (Courtesy of Hereaus Noblelight GmbH.)

in [2−2] cycloadditions.[52] The device is ideally suited to complement the drug discovery chemist's toolbox. Also Genzyme has used microflow reactors in combination with design of experiment techniques to optimize the photoisomerization of an intermediate of doxercalciferol (vitamin D_2).[53] Lek pharmaceuticals has disseminated a continuous-flow photochemical protocol for the benzylic bromination of a 5-methyl-substituted pyrimidine, which is an important intermediate in the synthesis of rosuvastatin.[54] The flow reactor consists of an FEP capillary (760 μm ID, 18 mL volume) coiled around a quartz cooling jacked and irradiated by a 150-W medium-pressure mercury lamp. In flow, 58.3-mmol/h product could be produced (583 g/day), which is four times higher than the corresponding batch process. Eli Lilly has collaborated with the Kappe group to develop a broad range of different photochemical transformations in flow. This includes benzylic brominations,[55] benzylic fluorinations,[56] α-trifluoromethylations of ketones[57] and CH arylations of arenes.[58] Novartis has worked together with Booker-Milburn *et al.* to develop continuous-flow protocols for photocycloadditions.[59] Janssen

Pharmaceuticals participate in the Photo4Future ITN consortium to introduce photoflow chemistry in their medicinal chemistry programs.[60]

9.5 Conclusion

As one can notice by reading this chapter, there are not too many examples of photochemical transformations in the industry. This has partly to do with the reluctance to implement photochemistry by organic chemists and process chemists. Due to lack of suitable technology and poor understanding of the underlying chemistry, photochemical transformations indeed were avoided as much as possible and were only pursued in the absence of any other alternative.

However, due to the recent success stories in photoredox catalysis and continuous-flow photochemistry, the scientific community is slowly changing its mind. The pharmaceutical industry embraces photoredox catalysis as a new key transformation to enable reactions which are otherwise elusive. Despite the highly secretive nature of this sector, publications and oral communications definitely show that the pharmaceutical industry will implement photochemistry in large-scale synthetic processes in the near future.

Nevertheless, moving forward is not without a challenge. The energy cost of irradiation, one of the main obstacles in the past, can now be dramatically lowered by energy-efficient and long-lived LEDs that produce near monochromatic light. However, the need for driver current controllers and the use of appropriate heat sinks to guarantee steady and reproducible optical output may complicate the irradiation setup. Moreover, the cost of UV LEDs is still very high and practically only available for >350 nm range, leaving traditional mercury lamps as source of choice for excitation at lower wavelengths.

With regard to scalability and reactor design, micro or millireactors can be employed in a numbering-up fashion, but many researchers in industry are not convinced that such devices are suited to meet the required productivity. Probably, a smart scaling strategy using a combination of larger diameter tubing and numbering-up approach could demonstrate the successful chemical manufacturing (e.g. of Active Pharmaceutical Ingredients) based on a photochemical transformation, thus convincing the industry of the scale-up potential of this technology.

In addition, continuous-flow photochemistry is a promising green chemistry and technology tool. Most photochemical transformations use less toxic reagents and can be performed at room temperature. In this respect, the combination of chemical synthesis with solar energy could represent the ultimate green process for the chemical industry.

References

1. W. A. Noyes and L. S. Kassel, *Chem. Rev.* **3**, 199 (1926).
2. K. H. Pfoertner, *J. Photochem. Photobiol. A* **51**, 81 (1990).
3. K. H. Pfoertner, *J. Photochem.* **25**, 91 (1984).
4. B. Koenig, *Chemical Photocatalysis* (de Gruyter, Berlin, 2013).
5. D. A. Nicewicz and D. W. C. MacMillan, *Science* **322**, 77 (2008).
6. N. A. Romero and D. A. Nicewicz, *Chem. Rev.* **116**, 10075 (2016).
7. K. L. Skubi, T. R. Blum and T. P. Yoon, *Chem. Rev.* **116**, 10035 (2016).
8. C. K. Prier, D. A. Rankic and D. W. C. MacMillan, *Chem. Rev.* **113**, 5322 (2013).
9. J. W. Tucker and C. R. J. Stephenson, *J. Org. Chem.* **77**, 1617 (2012).
10. J. M. R. Narayanam and C. R. J. Stephenson, *Chem. Soc. Rev.* **40**, 102 (2011).
11. D. Cambie, C. Bottecchia, N. J. W. Straathof, V. Hessel and T. Noel, *Chem. Rev.* **116**, 10276 (2016).
12. M. B. Plutschack, C. A. Correia, P. H. Seeberger and K. Gilmore, *Top. Organomet. Chem.* **57**, 43 (2016).
13. Y. Su, N. J. W. Straathof, V. Hessel and T. Noel, *Chem. Eur. J.* **20**, 10562 (2014).
14. Z. J. Garlets, J. D. Nguyen and C. R. J. Stephenson, *Isr. J. Chem.* **54**, 351 (2014).
15. T. Van Gerven, G. Mul, J. Moulijn and A. Stankiewicz, *Chem. Eng. Process.* **46**, 781 (2007).
16. Y. Su, K. Kuijpers, V. Hessel and T. Noel, *React. Chem. Eng.* **1**, 73 (2016).
17. K. Jaehnisch and U. Dingerdissen, *Chem. Eng. Technol.* **28**, 426 (2005).
18. N. M. Reis and G. Li Puma, *Chem. Commun.* **51**, 8414 (2015).
19. K. S. Elvira, R. C. R. Wootton, N. M. Reis, M. R. Mackley and A. J. deMello, *ACS Sust. Chem. Eng.* **1**, 209 (2013).
20. M. Oelgemoeller, *Chem. Rev.* (2016) **116**, 9664 (2016).

21. C. Jung, K.-H. Funken and J. Ortner, *Photochem. Photobiol. Sci.* **4**, 409 (2005).
22. J. M. Gordon, D. Babai and D. Feuermann, *Sol. Energy Mater. Sol. Cells* **95**, 951 (2011).
23. J. Blanco, S. Malato, P. Fernandez-Ibanez, D. Alarcon, W. Gernjak and M. I. Maldonado, *Renew. Sust. Energy Rev.* **13**, 1437 (2009).
24. M. N. Chong, B. Jin, C. W. K. Chow and C. Saint, *Water Res.* **44**, 2997 (2010).
25. H. Metzger, D. Fries, U. Heuschkel, K. Witte, E. Waidelich and G. Schmid, *Angew. Chem.* **71**, 229 (1959).
26. M. Fischer, *Angew. Chem. Int. Ed.* **17**, 16 (1978).
27. D. D. Bikle, *Chem. Biol.* **21**, 319 (2014).
28. http://www.marketsandmarkets.com/ (accessed on 18 April 2016).
29. M. A. Corsello and N. K. Garg, *Nat. Prod. Rep.* **32**, 359 (2015).
30. Y. Tu, *Nat. Med.* **17**, 1217 (2011).
31. J. Turconi, F. Griolet, R. Guevel, G. Oddon, R. Villa, A. Geatti, M. Hvala, K. Rossen, R. Goeller and A. Burgard, *Org. Process Res. Dev.* **18**, 417 (2014).
32. A. Burgard, T. Gieshoff, A. Peschl, D. Hoerstermann, C. Kelechovsky, R. Villa, S. Michelis and M. P. Feth, *Chem. Eng. J.* **294**, 83 (2016).
33. M. Peplow, *Nature* **530**, 389 (2016).
34. K. G. Fahlbusch, F. J. Hammerschmidt, J. Panten, W. Pickenhagen, D. Schatkowski, K. Bauer, D. Garbe and H. Surburg, in *Ullmann's Encyclopedia of Industrial Chemistry* (Wiley-VCH Verlag, 2003), DOI: 10.1002/14356007. a11_141.
35. W. Pickenhagen and D. Schatkowski, DE 19645922A1 (1998).
36. A. M. Braun, M.-T. Maurette and E. Oliveros, *Photochemical Technology* (Wiley, Chichester, 1991).
37. G. Ohloff, E. Klein and G. O. Schenck, *Angew. Chem.* **73**, 578 (1961).
38. G. O. Schenck, G. Ohloff and E. Klein, DE1137730 (1962).
39. D. Ravelli, S. Protti, P. Neri, M. Fagnoni and A. Albini, *Green Chem.* **13**, 1876 (2011).
40. M. Oelgemoeller, C. Jung and J. Mattay, *Pure Appl. Chem.* **79**, 1939 (2007).
41. N. Monnerie and J. Ortner, *J. Sol. Energy Eng.* **123**, 171 (2001).
42. M. Freund, R. Csikos, S. Keszthelyi and G. Mozes, *Paraffin Products — Properties, Technologies, Applications* (Elsevier Science, 1983).
43. M. R. V. Sahyun, *Kirk-Othmer Encyclopedia of Chemical Technology* (Wiley, 2000), DOI:10.1002/0471238961.1921182219010825.a01.
44. K. A. Lipper and E. Loeser, in *Ullmann's Encyclopedia of Industrial Chemistry* (Wiley-VCH Verlag, 2012), DOI:10.1002/14356007.o04_001.
45. Y. Manabe, Y. Kitawaki, M. Nagasaki, K. Fukase, H. Matsubara, Y. Hino, T. Fukuyama and I. Ryu, *Chem. Eur. J.* **20**, 12750 (2014).
46. H. Ehrich and D. Linke, *Chimia* **56**, 647 (2002).
47. H. Matsubara, Y. Hino, M. Tokizane and I. Ryu, *Chem. Eng. J.* **167**, 567 (2011).
48. T. H. Rehm, *Chem. Eng. Technol.* **39**, 66 (2016).

49. H. G. Yayla, F. Peng, I. K. Mangion, M. McLaughlin, L.-C. Campeau, I. W. Davies, D. A. DiRocco and R. R. Knowles, *Chem. Sci.* **7**, 2066 (2016).
50. S. D. Halperin, D. Kwon, M. Holmes, E. L. Regalado, L. C. Campeau, D. A. DiRocco and R. Britton, *Org. Lett.* **17**, 5200 (2015).
51. S. Werner, R. Seliger, H. Rauter and F. Wissmann, EP 2065387A2 (2008).
52. A. Vasudevan, C. Villamil, J. Trumbull, J. Olson, D. Sutherland, J. Pan and S. Djuric, *Tetrahedron Lett.* **51**, 4007 (2010).
53. B. G. Anderson, W. E. Bauta and W. R. Cantrell, *Org. Process Res. Dev.* **16**, 967 (2012).
54. D. Sterk, M. Jukic and Z. Casar, *Org. Process Res. Dev.* **17**, 145 (2013).
55. D. Cantillo, O. de Frutos, J. A. Rincon, C. Mateos and C. O. Kappe, *J. Org. Chem.* **79**, 223 (2014).
56. D. Cantillo, O. de Frutos, J. A. Rincon, C. Mateos and C. O. Kappe, *J. Org. Chem.* **79**, 8486 (2014).
57. D. Cantillo, O. de Frutos, J. A. Rincon, C. Mateos and C. O. Kappe, *Org. Lett.* **16**, 896 (2014).
58. D. Cantillo, C. Mateos, J. A. Rincon, O. de Frutos and C. O. Kappe, *Chem. Eur. J.* **21**, 12894 (2015).
59. L. D. Elliott, J. P. Knowles, P. J. Koovits, K. G. Maskill, M. J. Ralph, G. Lejeune, L. J. Edwards, R. I. Robinson, I. R. Clemens, B. Cox, D. D. Pascoe, G. Koch, M. Eberle, M. B. Berry, K. I. Booker-Milburn, *Chem. Eur. J.* **20**, 15226 (2014).
60. www.Photo4Future.com (accessed on 20 June 2016).

Index